SIR JAMES DEWAR, 1842–1923
A RUTHLESS CHEMIST

Science, Technology and Culture, 1700–1945

Series Editors

David M. Knight
University of Durham

and

Trevor Levere
University of Toronto

Science, Technology and Culture, 1700–1945 focuses on the social, cultural, industrial and economic contexts of science and technology from the 'scientific revolution' up to the Second World War. It explores the agricultural and industrial revolutions of the eighteenth century, the coffee-house culture of the Enlightenment, the spread of museums, botanic gardens and expositions in the nineteenth century, to the Franco-Prussian war of 1870, seen as a victory for German science. It also addresses the dependence of society on science and technology in the twentieth century.

Science, Technology and Culture, 1700–1945 addresses issues of the interaction of science, technology and culture in the period from 1700 to 1945, at the same time as including new research within the field of the history of science.

Also in the series

Popularizing Science and Technology in the European Periphery, 1800–2000
Edited by Faidra Papanelopoulou, Agustí Nieto-Galan and Enrique Perdiguero

Essays on David Hume, Medical Men and the Scottish Enlightenment
'Industry, Knowledge and Humanity'
Roger L. Emerson

The Language of Mineralogy
John Walker, Chemistry and the Edinburgh Medical School, 1750–1800
Matthew D. Eddy

Sir James Dewar, 1842–1923
A Ruthless Chemist

J.S. ROWLINSON
University of Oxford

Routledge
Taylor & Francis Group
LONDON AND NEW YORK

First published 2012 by Ashgate Publishing

2 Park Square, Milton Park, Abingdon, Oxon OX14 4RN
711 Third Avenue, New York, NY 10017, USA

Routledge is an imprint of the Taylor & Francis Group, an informa business

First issued in paperback 2017

British Library Cataloguing in Publication Data
Rowlinson, J.S.
Sir James Dewar, 1842–1923: A Ruthless Chemist.
– (Science, Technology and Culture, 1700–1945)
1. Dewar, James, 1842–1923. 2. Chemists – Great Britain – Biography. I. Title II. Series
540.9'2–dc23

Library of Congress Cataloging-in-Publication Data
Rowlinson, J.S. (John Shipley), 1926–
Sir James Dewar, 1842–1923: A Ruthless Chemist / J.S. Rowlinson.
p. cm. – (Science, Technology and Culture, 1700–1945)
Includes bibliographical references and indexes.
1. Dewar, James, 1842–1923. 2. Chemists – Great Britain – Biography. 3. Inventors – Great Britain – Biography. 4.Science – Great Britain – History – 19th century. 5. Science – Great Britain – History – 20th century. I. Title.
QD22.D345R69 2012
540.92 – dc23
[B] 2012005199

ISBN 978-1-4094-0613-6 (hbk)
ISBN 978-1-138-10931-5 (pbk)

JAMES DEWAR

Persons of good sense, I have since observ'd, seldom fall into it [professional disputation] except Lawyers, University Men, and Men of all Sorts that have been bred at Edinborough.

Benjamin Franklin, Autobiography, 1771

… an individual of untiring energy, united to courage and inventiveness of no common order, and indispensable for the furtherance of this important enterprise.

Agnes Clerke, Lecture on 'Low temperature research at the Royal Institution, 1893–1900', 1901

I know – as your Grace also knows – that Sir James Dewar is a man of quarrelsome disposition and ungovernable temper …

John W. Gordon, barrister-at-law, to the Duke of Northumberland, President of the Royal Institution, 1914

… that crusty dreamer, who loved poetry, and made and played on his fiddle; who studied the sky at night through a skylight in the roof of the RI at the age of 80, invented an explosive, and treasured a soap bubble.

Gwendy Caroe, daughter of William Bragg, in *The Royal Institution: An Informal History*, 1985

Portrait of James Dewar in 1902 holding a flask in his laboratories of
The Royal Institution, (photogravure) by Alexander Scott.
Courtesy of The Royal Institution, London/The Bridgeman Art Library.

Contents

List of Figures

Preface

The nineteenth century marked the coming of age of the physical sciences. In 1800, astronomy was the only observational science to have reached maturity, as exemplified by the appearance a few years later of the five volumes of Laplace's *Mécanique Céleste*. This book has been characterised both as the culmination of the Newtonian era of the eighteenth century and the first recognisable book of theoretical physics of the nineteenth. By 1800 the elasticity of solids, the elasticity and patterns of flow of gases and liquids, and simple optics and electrostatics had all been studied and reduced to mathematically ordered descriptions, but little progress had been made with most of the other subjects that we now classify as physics. The situation in chemistry was no better; many facts had been established but the subject still had more in common with the classificatory field of natural history than with what we now see as a physical science. By 1900 the situation had changed totally. The powerful experimental and theoretical structure of classical physics was almost complete, and chemistry, following the acceptance of Dalton's atoms, had become a coherent field of experimental science. Organic chemistry, a subject that did not exist in 1800, became one of the great achievements of the time. The devising of systematic methods of analysis and synthesis had created an immense corpus of knowledge that must be reckoned one of the glories of the intellectual life of the nineteenth century. Chemistry's link with the biological sciences was, however, only just starting in 1900. Its links with physics had been underway since the middle of the nineteenth century. At first physical chemistry was most successful in those aspects of chemistry, such as the thermodynamics of solutions and the kinetics of reactions, which made little immediate demand on atomic interpretations. Spectroscopy was valued principally as an analytical tool that proved its worth in the discovery of new elements, particularly with the inert gases at the end of the century. But the link of chemistry to physics could not reach fruition until the advent of quantum mechanics in the twentieth century. Only then could it be understood how the physics of atoms differed fundamentally from the physics that sufficed for billiard balls.

Those behind this huge body of nineteenth-century science have received their due attention from historians, but the most studied have been those whose theoretical ideas drove the subjects forward. The men responsible for the experimental foundations of this theoretical structure have been less studied, with the exception of perhaps the greatest of them all, Michael Faraday. This book is an account of the life of one of them, James Dewar, a Scots chemist by training whose principal work lay in fields that we now reckon to be physics – atomic spectroscopy, the liquefaction of gases, and so the achievement in the laboratory of low temperatures, which turned out to be an environment in which the properties of matter differ in unexpected ways from those at ambient temperatures. His experimental skills were comparable with those of Faraday but he never showed the same interest in the theory of what he had discovered. He was a man of independent thought and an exceptionally wide range of interests, which he followed with a ruthless single-minded effort that has rarely been matched. He had the benefit of skilled assistants, but he never had any students and he created no 'school'. He worked mainly in London and, although a man of few friends and many enemies, was at the centre of the British scientific establishment for 30 years. He left behind a mass of papers and correspondence but he has been little studied by historians, perhaps, in part, because of his unsympathetic character, and perhaps because his dreadful handwriting makes much of what he has left almost inaccessible. His life and times should, however, not be neglected because of such adventitious problems and this book is an attempt to fill a gap in our knowledge of British chemistry and physics in the last quarter of the nineteenth century.

Dewar held chairs at both Cambridge and at the Royal Institution in London but so much of his important research was carried out at the RI, as the Institution is commonly called, that writing his history could lead to an attempt to write the history of the RI also. I have tried to avoid that outcome which has, fortunately, now been taken care of in the recent book edited by Frank James on *'The Common Purposes of Life': Science and Society at the Royal Institution of Great Britain* (Ashgate, 2002). This book is described in the Notes and References simply as *James*.

The wide range of Dewar's research and his ability to work hard in more than one field at any time make it impossible to write a chronological account of his life. The book is therefore organised to follow his career according to the places where he worked and the subjects that were at the centre of his research. A chronological summary at the end of the book will, I hope, allow the reader to see how these subjects interlock in time.

He wrote no books, and I have not included a list of his papers since two sets have already been published. The more useful, and more reliable, are those published by Cambridge University Press: the *Collected Spectroscopic Papers* with George Liveing (1915), and the two volumes of other *Collected Papers*, published posthumously (1927) – see List of Abbreviations. The second list is *A Record of the Scientific Work of Sir James Dewar*, complied by Henry Young, the Assistant Secretary and Librarian from 1889 to 1928, which was printed by the Chiswick Press in London and published privately and anonymously in 1933. Young lists the papers by title only, but he does, however, record also many of the journals that reprinted or abstracted Dewar's articles from their primary sources and has a few minor contributions that the *Collected Papers* omit. Both lists ignore a polemical pamphlet that Dewar issued in 1891, and most of his minor contributions to the many squabbles of his day. More surprisingly, the *Collected Papers* omit without a mention the long article on 'Liquid Gases' that he published in the 10th and 11th editions of *Encyclopaedia Britannica* in 1902 and 1911. Perhaps there was a problem over copyright, since Joseph Larmor's article on the theory of radiation in the 10th edition is also missing from his *Mathematical and Physical Papers*. Both the *Collected Papers* and Young's list overlook also a substantial but unsigned article on 'Alum' that he wrote for the 9th edition of the *Encyclopaedia* in 1875.

The liquefaction of hydrogen, his invention of the vacuum flask, and the study of the properties of matter at low temperature are at the centre of Dewar's claim to fame. The controversies that surrounded some of this work cannot be fully appreciated without an understanding of the physics of gas liquefaction. This is a tricky topic that he and many of his contemporaries got wrong, and, indeed, which is often misunderstood to this day. I have tried to give a short and simplified account of the principles in an appendix. I hope, however, that the most important points at issue can be understood in outline from the main text alone.

For their help and advice, I thank Kay Baxandall and Kenneth McRae (Ruhemann family), Bill Brock (Dewar-Armstrong correspondence), Mark Child (decipherment of Dewar's handwriting), Alwyn Davies (Ramsay), John Eland (spectroscopy), John Freeman (diagrams), Kostas Gavroglu (Dewar-Kamerlingh Onnes correspondence), Peter Harman (Maxwell), Roland Jackson (Athenaeum Club), Sarah Kellam (Ansdell family), David Knight (Hodgkins bequest and editorial advice), Roger Lovett (Peterhouse), Keith McLauchlan (Kamerlingh Onnes), Sy Mauskopf (Dewar–Nobel correspondence), Johnathon Orr-Ewing (Carlton Clubs in London), Adam Perkins (Cambridge University Archives), Sami Timini (Asperger's syndrome), Philippa Wright (Reynolds family), and my late wife for her reading of the draft manuscript, to its great benefit. Oxford's libraries contain much of the material I needed, but it is scattered over many sites, of which I have used nine. My greatest debt here has been to the staff of the Radcliffe Science Library who have gone to great trouble, at a difficult time of reorganisation, to provide every service that I could desire. In Cambridge I have used the University Library and in London the archives of the Royal Society and of the Royal Society of Chemistry. The bulk of Dewar's papers, however, are in the Royal Institution and here the Archivist, Jane Harrison, and the Professor of the History of Science, Frank James, have generously given me hours of their time to help me to navigate the immense resources of the Institution's archives. For permission to quote from documents in their possession I thank the Syndics of Cambridge University Library, Imperial College, London, the Museum of the History of Science, Oxford, the Royal Institution of Great Britain, the Royal Society of Chemistry, and the Royal Society of London.

J.S. Rowlinson
Oxford, January 2012

Nomenclature and Units

The naming of chemical compounds and of physical units has changed over the years, but for about 100 years from the middle of the nineteenth century it was relatively stable, and I use, as far as is convenient, the names and units of that time. Thus C_2H_4 is 'ethylene', not the obsolete 'olefiant gas', nor the current 'ethene'. The element S is written here as 'sulphur', not the current and more defensible 'sulfur'. For the 'inert' gases, see Chapter 9.

The use of British units such as the inch, pound mass, and degree Fahrenheit had, by 1870, been abandoned by most academic British scientists but was still customary among engineers. Here metric units and those related to them are used whenever posssible: lengths in metres and its sub-units, mass in grammes or kilogrammes, time in seconds, and temperature in kelvin (see below). The common unit of volume used in laboratory work is the cubic centimetre, which is properly written cm^3, but for which I have often substituted the colloquialism, cc, as was the custom then and now.

Pressure presents problems. The modern fundamental unit is the pascal (Pa), or newton per square metre, but this is impossibly small for most purposes. Fortunately 10^5 Pa is 1 bar which is close to atmospheric pressure, where 1 atm is now defined as 1.01325 bar. These two names are often interchangeable for rough calculations, and the atmosphere is still the name of choice in casual conversation. For vacuum work the common unit was the pressure from a column of mercury 1 mm high, now conventionally fixed by choosing 760 mmHg to be equal to 1 atm, or 1 mmHg to be 133.322 Pa.

Temperatures below the ice-point were, in the nineteenth century, usually described by negative numbers in degrees 'centigrade' (now 'celsius'). Thus the normal boiling temperatures of liquid oxygen and liquid hydrogen are, in round figures, -183 °C and -259 °C. The relation between these temperatures is, however, made clearer by expressing them as degrees above the absolute zero of the thermodynamic scale (-273 °C), that is, as 90 K and 14 K respectively, where it is convenient to use the modern symbol of K for the unit 'kelvin', instead of the older 'centigrade degree absolute', with no accepted symbol. Energy, whether mechanical or thermal, can best be expressed in joules, where a joule (J) is a newton metre (Nm). It was, however, conventional until recently to use the joule for mechanical energy and the calorie for thermal energy, where a calorie is the energy needed to raise the temperature of 1 g of water by 1 K, a definition that does not fix the size of the calorie exactly, but where the unit most commonly used by scientists, the 'thermochemical calorie', is defined to be 4.184 J.

The value of money changes over time and its buying power cannot be defined precisely because of the change in the nature of the goods and services bought over the years. But rough estimates of the buying power of £100 in 2010 are that it matched that of £2.25 in 1880, that of £2 in 1900, and, after the inflation of the First World War, of £4 in 1920. The 'real value' of wages, that is, of their buying power in current terms, roughly doubled over the 40 years.

Abbreviations

Dewar's handwriting is often illegible and difficulties in quotations from his letters and notebooks are marked in one of three ways: [was] denotes a word that is not in his text but which seems to be needed to make sense of the passage; [?] means that a word (or words) is illegible and so has been omitted; [inner?] means that the word is probably 'inner'.

BAAS	British Association for the Advancement of Science (British Science Association since 2009)
BJHS	*British Journal for the History of Science*
BMFRS	*Biographical Memoirs of Fellows of the Royal Society*
CN	*Chemical News*
CP	Lady Dewar, J.D. Hamilton Dickson, H.M. Ross and E.C. Scott Dickson, *Collected Papers of Sir James Dewar* (CUP, 1927) 2 v.
CPS	G. Liveing and J. Dewar, *Collected Papers on Spectroscopy* (CUP, 1915)
CR	*Comptes Rendus de l'Académie des Sciences, Paris*
CU	Cambridge University
CUA	Cambridge University Archives
CUL	Cambridge University Library
CUP	Cambridge University Press
CUR	Cambridge University Archives, Registry File
DFRL	Davy Faraday Research Laboratory
DSB	C.C. Gillispie, ed., *Dictionary of Scientic Biography* (Scribners, New York, 1970–81) 18 v.
James	F.A.J.L. James, ed., *'The Common Purposes of Life': Science and Society at the Royal Institution of Great Britain* (Ashgate, Aldershot, 2002)
JCS	*Journal of the Chemical Society*
JCS Trans	*Transactions of the Journal of the Chemical Society*
NRRS	*Notes and Records of the Royal Society*
ODNB	*Oxford Dictionary of National Biography* (OUP, 2004) 60 v.
ONFRS	*Obituary Notices of Fellows of the Royal Society*
OUP	Oxford University Press
PCS	*Proceedings of the Chemical Society*
PM	*Philosophical Magazine*

PRI	*Notices of the Proceedings of the Meetings of the Members of the Royal Institution*
PRS	*Proceedings of the Royal Society* [of London], divided into two parts in 1904, *A* for the physical sciences and *B* for the biological
PRSE	*Proceedings of the Royal Society of Edinburgh*
PTRS	*Philosophical Transactions of the Royal Society* [of London], divided into two parts in 1887, *A* for the physical sciences and *B* for the biological
Rep. BAAS	*Reports of the British Association for the Advancement of Science*. Until 1879 the *Trans. Sec.*were paginated separately. The year shown for each volume is that of the meeting itself; publication was a year later
RI	Royal Institution of Great Britain (I see no merit in the current abbreviation, Ri)
RIA	Archives of the Royal Institution
RIA D, DA, DB etc.	Dewar papers in the *RIA*
RIA MC DFRL	Minutes of the Committee of the Davy Faraday Research Laboratory
RIA MM	Managers' Minutes of the RI. References are by the date of the meeting only. The minutes up to 1903, originally bound in 15 volumes, were reproduced in facsimile in seven volumes and published by Scolar Press, Ilkley, Yorkshire in 1977. The 7th volume covers the years 1874–1903 and is the one used here. From 1903 onwards the original records at the RI have been used.
RS	Royal Society [of London]
RSA	Archives of the Royal Society
RSC	Royal Society of Chemistry
TRSE	*Transactions of the Royal Society of Edinburgh*

Titles of other less frequently quoted journals are given in a generally recognised form.

Chapter 1
Boyhood

The original Dewar was a pilgrim or wanderer, deoradh in Gaelic, a word in which the last two consonants are silent.[1] He cannot, however, have wandered far, for throughout the second half of the nineteenth century the name was restricted largely to the central belt of Scotland, that is, to the triangle formed by Glasgow, Edinburgh and Perth, with the greatest number always being in the county of Perthshire.[2] By then the most widely-known member of the tribe was John Dewar who had set up his whisky business in Perth in 1846. The James Dewar of this book was, as far as is known, no direct relation of John, although his family was in a similar line of business.

Thomas Dewar, of Overtoun, near Dumbarton on the Clyde, moved in the 1760s to Kincardine-on-Forth.[3] This was then a small fishing and trading town where he established or took over the Unicorn Inn on Excise Street.[4] The business flourished and his grandson, also called Thomas, who described himself as vintner and innkeeper, married Ann (or Anna or Agnes) Eadie, the daughter of Hugh Eadie, a local shipbuilder – reportedly 'a very charming and clever' lady.[5] They were to have seven sons, but in 1841, just before James was born, the family was not living together. In the census for that year the three eldest boys, Thomas (13), Ebenezer (10) and Robert (7) were living nearby in Kilbagie Street with Margaret Dewar (30), a dressmaker, and presumably the younger sister of Thomas, the father. At the Unicorn Inn were the parents, both 35, with Alex[ander] (6), Hugh (2) and John (8 months). Nothing more is heard of John and it may be presumed that he was sickly, that the trouble of caring for him had led to the temporary division of the family, and that he died some time after the census had been compiled. The next year James, the youngest son, was born on 20 September 1842. There is now a Dewar Avenue on the north side of the town of Kincardine, named in his honour.

He was brought up according to the beliefs of the Auld Licht, a conservative branch of the established Presbyterian Church in which his father was on the local Board of Management. He went to the New Subscription School, a non-denominational school where the level of teaching varied according to the abilities of the one teacher. At first this was a Mr Hogg but from 1855 it was for three years his elder brother Alexander. Under his guidance the school flourished and was commended in the summer of that year at an inspection by the local Presbytery. Unfortunately for James, Alexander left in 1858 to go to Edinburgh to study medicine. His successor, a Mr Dow of Culross, was not a success, but James left the school that year before his education could suffer much. His family life was blighted by two tragedies. In March 1852, when he was nine, his mother died. At around this time he was soaked by a fall through the ice, after which he wandered about outside for some time in his wet clothes until they were dry, so that his family might not know of the accident.[6] As a result of this escapade he contracted a rheumatic fever, and was forced to go around on crutches for two years. He retained a limp for the rest of his life. The fever forced him to give up playing the flute, at which he was becoming adept, but his father engaged a local fiddler who taught him to play his instrument. His skills were enhanced when a joiner then showed him how to build a violin – surely a formidable achievement for a boy of 12? One of his instruments, marked *James Dewar – 1854*, survived to be played at the celebration of the jubilee of his wedding in 1921, and again after Kurt Mendelssohn's Friday Evening Discourse at the Royal Institution in 1966; it is still there. It is, however, surprising that the dexterity he showed in making this instrument, and later in life, in his experiments, did not result in more legible handwriting.

His father was a man of unusual abilities. He was a keen naturalist who therefore catered for the annual dinner of the local Horticultural Society, and of many other of the town's organisations. More remarkably, when he realised that the scale of his operations was restricted by his facilities, he constructed a plant to generate gas with which he illuminated his inn, or 'Dewar's Hotel' as it was now often called. A few years later the Kincardine Light and Gas Company was set up and he transferred his equipment to their services, taking care, however, to become also a shareholder in the Company. The father's business sense and engineering skills, and the tuition from the fiddle-maker, showed James the importance of cultivating abilities and tastes beyond those of the academic learning he was taught at school; it is not too fanciful to see his whole career so presaged in his early experiences.

The death of his mother was followed by the death of his father on 2 September 1857, a few days before James's 15th birthday. None of the brothers was interested in or able to continue the business which was therefore put on the market at the end of the month. His father had appointed as trustees: his brother Alexander, a wine and spirit merchant in Leith; the local minister, Andrew Gardiner; and a lawyer, Ebenezer Mill. His estate was substantial; it comprised the Inn, some other properties in Kincardine, his shares in the Gas Company, the crops of several fields, and £1657 in cash. It was divided equally between the six sons, with those who were still minors having their share administered by the trustees. Gardiner, himself a classical scholar, was able to persuade his colleagues that James was clearly an able lad on whom they should spend the money to send him as a boarder to Dollar Institution (now Dollar Academy). This was a fee-paying school in the village of that name, some eight miles from Kincardine, where he was at last able to receive an education that his abilities justified. Meanwhile, when not at school, he lived with his then unmarried brother Robert who had set up shop as a draper in the town. At Dollar Institution he boarded with one of the masters, a Dr Lindsay, who was a fine mathematician. He and James found an old sundial in the garden which they restored and calculated the right orientation for its remounting. When Dewar returned to speak at the school in 1907 he said of him: 'It was entirely due to his influence and under his direction that the bent of my life was directed toward the side that it has been, and probably the only side where I should have succeeded.' Lindsay also taught chemistry although without the benefit of a laboratory. Naturally the school taught the classical languages that were then held to be the foundation of a gentleman's education, but it also offered a wider range of modern languages than most schools of the day. It was, however, on the mathematical and scientific side that James shone. When he left the school in August 1859 he received the first prizes for geometry, experimental philosophy, chemistry, and mineralogy and geology; second prizes for algebra and human physiology; and the third prize for mechanical drawing. He missed a share in the mathematical medal since the rules required that the pupil who received it must have attended the school for two years and James had been there only for one. The external examiner, Philip Kelland, the Professor of Mathematics at Edinburgh, singled out James in his report as 'one lad who had scarcely any knowledge of mathematics, yet, through only one session at the school he showed abilities that might have been the labour of three years'.

Chapter 2
Edinburgh

James Dewar matriculated at the University of Edinburgh for the start of the session in the autumn of 1859. His brother Alexander had started on his medical course and they took lodgings with two elderly ladies. James stayed with them until his marriage at Bristo United Presbyterian Church on 8 August 1871 to Helen Rose Banks, the eldest surviving daughter, born in 1849, of the late William Banks, an Edinburgh engraver and printer.[1] James had cut short his attendance at the British Association for the Advancement of Science, which met in Edinburgh that year from 2 to 9 August, in order to get to his wedding on time. Here, after the ceremony, he played on his juvenile fiddle. The married couple made their home at 15 Gilmour Place.

The University[2] was then an institution run by the city of Edinburgh which offered degrees in Arts, Law, Divinity and Medicine. The last was a necessary qualification for an aspiring medical man, but any other degree was an expensive luxury for most of the students. In the year that Dewar entered the University, 42 men received the MD, but in all other subjects only 25 took the MA. Most men simply attended those professors' lectures in which they were interested and obtained a certificate at the end of the course which vouched for their attendance, for their performance in any tests they had taken, and for any prizes or medals that they had been awarded. When Dewar arrived the Professor of Natural Philosophy was James David Forbes with whom he worked briefly as a laboratory assistant. Forbes left that year to become Principal of the University of St Andrews, but he gave to Dewar the travelling chemical case that he taken on his Alpine travels. He was succeeded by [Peter] Guthrie Tait,[3] the Professor of Mathematics at Queen's College, Belfast. Tait had published little at that time but the city fathers were looking for a good teacher and chose him in preference to James Clerk Maxwell,[4] the Professor at Marischal College, Aberdeen, who was to prove the more brilliant man but was not so good a lecturer. The Professor of Chemistry was also a new arrival; Lyon Playfair[5] was a Scots chemist who had made his reputation in the last 10 years as a public servant in London, notably when he was a leading organiser of the Great Exhibition of 1851. His return to academic chemistry was a surprise to many and he used it not to promote any research but to improve the organisation of science teaching in the University and to emphasise there the importance of chemistry to British industry. This was a time when industrial developments were first starting to depend directly on scientific advances, particularly in chemistry and electricity, and Playfair was one of those who were conscious that Britain was falling behind its Continental competitors in the extent and depth of its scientific education.

He wrote: 'Before entering upon my duties as a professor, a considerable sum had to be spent in equipping the laboratories and chemical museum with the full appliances for teaching.' He did not say that the 'considerable sum' came from his own salary for the first two years. Chemistry was a necessary subject for the medical students; it was they who accounted for the large numbers at Playfair's lectures. In his programme of reform he was helped by the passing of the Universities (Scotland) Act of 1858. This provided for a body called the Curators of Patronage to replace the Town Council as the appointing body for university professors, although the Council was still represented by four members of the seven-man board. The Act also provided for the appointment of paid assistants to those professors who needed to run practical classes. Playfair described the position when he arrived as: 'The old tradition of classes prevailed. The number of students was too large in each class for effective teaching. I introduced a tutorial system, like that of *répétiteurs* in France. The tutors followed my course of lectures, and drilled the students in practical exercises upon the subjects which I had discussed.' These posts replaced those of the privately-paid assistants that some professors, including Playfair, had taken on previously. Perhaps Playfair's most far-reaching university reform was his persuading the Senate to introduce a science degree in 1864, but it was some years before it became popular; two students qualified for the B.Sc. in 1866. From 1863 there was also a post of Extra-Academical Lecturer in chemistry which was occupied by Alexander Crum Brown (1838–1922),[2, 6] an Edinburgh medical graduate but also the first holder of a London University D.Sc.

Dewar attended the classes of Tait and Playfair, and the mathematics classes of Kelland, obtaining prizes from Tait and Kelland in 1860 and 1861.[7] His inclination was at first more to physics than to chemistry but, as Cargill Knott wrote, 'There was then no physical laboratory, and I have heard Tait lament that he was unable to make use of Dewar's ability in those very early days. Tait was also in the habit of saying that one of the greatest services he did to experimental science was in recommending Dewar to Lyon Playfair as demonstrator and assistant'.[3] The chemistry chair was then in the Faculty of Medicine and for a time it seems that Dewar thought of following his brother into a medical career, but he decided against it and finally threw his lot in with the chemists. He held the post of assistant to the Professor of Chemistry, under Playfair, until Playfair's resignation in 1869 to become the Member of Parliament for the Universities of Edinburgh and St Andrews. He was succeeded by Crum Brown, whom Dewar served until 1874.[2, 6] He was paid a salary of £100 a year from the University. His colleague John Young Buchanan,[8] who was later to be the chemist on the *Challenger* oceanic expedition, was assigned to those students who were not intending to follow a medical career, and Dewar to the more numerous medical men. He carried out his duties well; he was, a former student wrote, 'brimful of energy and enthusiasm, which he communicated to his class – always lambent, sometimes flashing brilliantly'. Meanwhile his brother Alexander had acquired a similar assistantship under James Syme, the Professor of Clinical Surgery.

A leading British industry of the day was the spinning, weaving and printing of cotton goods. Glasgow was then the second city of the kingdom and the central belt of Scotland was a thriving centre of these and many other industries. The Scottish universities produced a high proportion of the scientific and technically trained men and throughout Dewar's career many of those he worked with, or argued against, were the products of Scotland's universities and technical colleges. Playfair, who soon became convinced of Dewar's ability, recommended that he and Peter Griess, a chemist at a brewery, seek employment with the dyestuffs firm of Simpson, Maude and Nicholson, the proprietors of which (and Griess himself) had all trained at the Royal College of Chemistry in London. Nicholson, never a fit man, finally retired in 1868.[9] The salaries they then offered to Dewar and Griess were derisory and neither accepted the post. Dewar, moreover, now had other aims and began casting around for a topic of research.

The 1860s were a time when the chemical confusion of the previous decade was being resolved, with the increasing adoption of the 'new' atomic weights of C = 12, O = 16 etc., after the Karlsruhe Conference of 1860, which Crum Brown had attended at a time when he was preparing his thesis. The tetravalency of carbon was soon accepted, but the tetrahedral distribution of these valencies, although discussed in these years, had to wait until 1874 for a convincing defence. The linking of chains of carbon atoms in organic molecules was advocated by Archibald Scott Couper and August Kekulé from 1857–8 onwards. Couper was a Scots chemist who became Playfair's assistant in December 1858 at a time when Crum Brown was finishing as a student, although there is no evidence of any meaningful discussion between them.[10] Couper fell ill soon after his appointment and withdrew from active chemistry thus unfortunately leaving all the credit to go to Kekulé.[11]

It is unlikely that Dewar received much enlightenment from Playfair on the rapid unfolding of these new ideas although he always held him in high esteem. Playfair had given a lecture, 'On the constitution of salts' to the Chemical Society on 7 May 1863, but it had an old-fashioned air about it and he was still using the old atomic weights.[12] A more likely source of inspiration for Dewar was the work of Crum Brown who had come back to Edinburgh in 1863. His Edinburgh MD thesis of 1861 was on the unusual topic of the theory of chemical combination.[13] It was not printed at the time but much of its contents appeared in a paper[14] of 1865. In it he introduced the graphic representation of molecules by single or pairs of bonds between each of the atoms, in the way that now seems so obvious to us. Kekulé had proposed a different way of representing the linking the atoms and Crum Brown wrote in his thesis: 'Kekulé's graphic method is a most artificial one, and, I think we may say, certainly does not represent the actual arrangements of the atoms'; and of his own representation: 'Of course, I do not intend it to be supposed that this represents correctly, or even more correctly than Kekulé's method does, the actual arrangements of the atoms, but it is at least as probable; and all I wish to show is, that his is not the only possible arrangement.' In his paper he went a bit farther in his claim by adding a footnote to say: 'I do not mean to indicate the physical, but merely the chemical positions of the atoms.' He was aiming to represent the topology of the molecules correctly but not their geometry, as was probably impossible on a plane sheet of paper. Either symbolism can be used for straight-chain aliphatic molecules but Kekulé's led to ambiguities if the carbon chain is branched. In the same year, 1861, Alexander Butlerov of Kazan gave a paper on 19 September at the 36th Meeting of the Deutscher Naturforscher und Aertze in Speyer[15] in which he laid out ideas similar to those of Crum Brown, and who also insisted that he was not claiming geometric accuracy. His influence, if any, on Dewar was probably less than that of Brown who lectured on the theory of chemistry from 1863.[10] Slowly, however, the idea that these structural formulae had some geometrical reality crept in and it became reinforced by Kekulé's proposal of a hexagonal structure for benzene in 1865 and his first inkling of a tetrahedral carbon atom in the aliphatic series from around 1867.

Dewar's first paper arose from an attempt to learn more about the structure of benzene. He decided that its oxidation would break it into fragments from which he could infer more about the original structure. Unfortunately benzene remained unchanged on heating to 150 °C with potassium permanganate, so he tried instead to oxidise 'phenyl alcohol' which reacted rapidly with varying amounts of permanganate. The resulting yields of oxalic acid and other products confirmed for him that Kekulé's structure was correct, but he went on then to speculate on what other planar structures were compatible with the empirical formula C_6H_6. He constructed seven model structures from narrow strips of copper sheet, to represent the bonds, and nuts with the four radiating bonds, to represent the carbon atoms. His first paper was submitted by Playfair to the Royal Society of Edinburgh[16, 17] and appeared in its *Proceedings* for 1867. The models included both Kekulé's hexagon with its alternating single and double bonds and a hexagon with a bond across the ring connecting atoms 1 and 4. He made no comment on this structure and there is no evidence that he seriously considered it as an alternative to Kekulé's, but it soon acquired the name of 'Dewar benzene'. The other five structures were ones that we can now see were most unlikely (Figure 2.1) but, again, he was not advocating them, but simply showing what was topologically possible. He does not claim that his list is exhaustive for tetravalent carbon atoms. It does not, for example, include the 'centric' structure suggested by Adolph Claus in the same year, which has three bonds crossing the hexagon linking each of the three opposing pairs of carbon atoms, nor does it include possible structures with triple carbon–carbon bonds, nor non-planar structures such as the prismatic molecule suggested by Albert Ladenburg in 1869.

In spite of Dewar's apparent lack of interest in his sixth structure and his support for the second, or Kekulé, structure, the name of 'Dewar benzene' has become accepted and it is now too late to change it. He did, however, use this structure many years later for the bond arrangement of the middle ring in anthracene.[18] At first glance it seems to a modern eye that 'Dewar benzene' represents a substance that could not exist in the real world. It is, on a resonance picture, merely one of the 'valence bond' structures that enter into a complete quantal description of the planar hexagonally symmetrical ring. This view, however, is not correct, for 'Dewar benzene' is not a

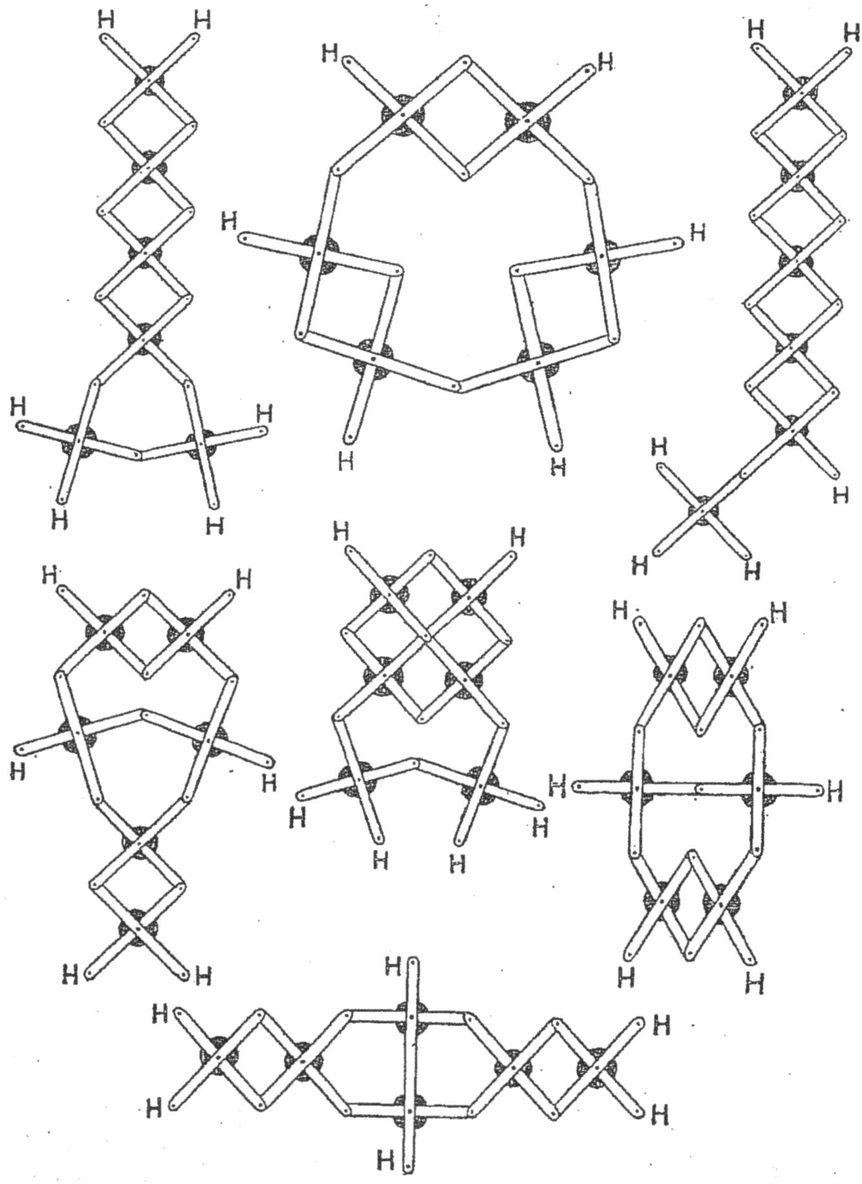

Figure 2.1 Dewar's models of possible forms of benzene, C_6H_6. These include Kekulé's structure which became conventional in the nineteenth century, the central structure in the first row, and the form called 'Dewar benzene', the right-hand structure in the second row.

planar molecule but one that is 'creased' along the 1-4 bond so the two wings meet at an angle of less than 180 degrees. Nevertheless such a molecular structure is difficult to prepare and is unstable with respect to the conventional benzene molecule. Three substituents in the 1,2,5-positions help to stabilise the structure by steric interaction and such a molecule was prepared in 1962, to be followed by the true Dewar benzene, but only in solution in pyridine, in 1963.[19]

Playfair sent a copy of Dewar's mechanical model to Kekulé in Ghent saying that Dewar was interested in the work that Kekulé was doing. An immediate invitation to come to Ghent led to Dewar spending the summer there in 1867.[20] David Brewster, who had succeeded Forbes as Principal of St Andrews University, wrote introductions for him so that he could meet Joseph Plateau and Adolphe Quetelet on his way through Brussels.[21] Dewar took with him to Ghent some of the coal-tar bases that he had been working on in Edinburgh, including picoline (one or more of the three isomeric methyl-pyridines). The work that he undertook there was a continuation of his first paper on trying to determine the structure of aromatic compounds from the products of their oxidation. It formed one of a group of miscellaneous papers, inconclusive in their results, and often short in length, that marked his first attempts to mark out his own line of research. They ranged over a wide range of organic, inorganic and physical topics but do not need detailed analysis here.

Whilst in Ghent Dewar met a man of Italian and German parentage – Guglielmo Koerner or Wilhelm Körner[22] – with whom he was soon on close terms. Körner was three years older than Dewar and a great practical joker. He and the 'wild young Scot', as Armstrong called him,[1] became friends and discussed together their problems in the field of aromatic chemistry, a field in which Körner was later to become prominent. One problem was the structure of pyridine, C_5H_5N. Dewar was to claim privately that, even before he went to Ghent, he had had the idea that it too has a hexagonal ring, one in which a trivalent radical CH of benzene is replaced by N. He did not, however, publish the idea until he put it into a paper that he read before the Royal Society of Edinburgh in June 1870, a paper that appeared in print only in the volume dated 1872, although a report appeared in *Chemical News* on 27 January 1871, which, in turn, was abstracted by Carl Schorlemmer for the Chemical Society.[23] Meanwhile Körner had tried to insert the same idea as a footnote in a paper that he published in April 1869 in the *Comptes rendus* of the French Academy, but which the editors had presumably thought too speculative and had removed.[24] Körner left Ghent at the same time as Dewar had returned to Edinburgh. He went to Cannizzaro's laboratory at Palermo and it was there that he succeeding in publishing his idea in an obscure local journal on 15 April 1869.[25] Dewar's claim to have been the first to suggest this structure is supported in a letter he wrote to Kekulé on 1 June 1869, saying that his work on picoline had been held up by an accident to his knee. He wrote: 'You recollect my idea in Ghent was that Picolin was Benzol in which the N functions as a triatomic element in the ring … If you don't remember Koerner does as lately I find this idea (nearly) "que je m'avais faite" to the Palermo Academy … This is a very small matter but coming from a friend, I don't like it'.[26] The French phrase is from Körner's Palermo paper and its use here shows that Dewar must have received a copy from Körner within six weeks of its publication. Dewar never took the matter further and stayed on good terms with Körner for the rest of his life, even seeing that his name was included in the list of Honorary Members of the Royal Institution elected at its centenary.[27] An older Dewar might not have been so forgiving about what may have been no more than a mis-remembering by Körner of informal discussions two years previously. Dewar brought back from Ghent either the idea, or the model itself, of a benzene structure that Kekulé had devised from balls and wires; it was essentially a three-dimensional representation of Crum Brown's type of structure in which the bonding lines emanating from each carbon atom are tetrahedrally disposed, then bent to form a carbon hexagon with alternate single and double bonds. In 1868 Dewar took this to the Norwich meeting of the British Association,[17] which he then joined, but only in the category of members who do not wish to buy the Annual Report.

In 1869 Dewar acquired a second post.[28] From 1820 William Dick had given lectures on veterinary science in Edinburgh and in 1833 had built himself a college in Clyde Street in which there was a dissecting room, accommodation both for overseas students and for sick animals, and a chemistry laboratory. He died in 1866 and bequeathed the endowed college to a trust managed by the Lord Provost and magistrates of Edinburgh. It later acquired the clumsy title of the Royal (Dick) Veterinary College and is now part of the University of Edinburgh. A Principal was chosen and in 1869 two new members of staff were appointed, an anatomist, who proved to be incompetent, and Dewar as a chemist. His title of Lecturer was soon advanced to that of Professor. He found the laboratory 'deficient in many respects' and spent £44/14/10 [£44.74] on improvements, telling the trustees who queried the amount that 'It seems ridiculous for anyone possessed of average knowledge to imagine that getting a table along with gas and water could have anything to do with analysis or investigation'. At the age of 27 the new professor had started his independent career in the way that he would continue.[29]

One day, when in the office of the Principal of the Veterinary College, he introduced himself to a man who was there seeking the bone of a horse; 'You and I should know each other, Dr McKendrick'. This was the physiologist, John Gray McKendrick,[30] a man a year Dewar's senior who, like Dewar, had lost both parents as a child but who had grown up in much poorer circumstances. He became Professor of Physiology at Glasgow in 1876. It was Dewar who suggested that they collaborate in studying the effect of light on the electrical properties of the eye. They borrowed a sensitive galvanometer from Tait which was of the pattern invented by Tait's friend, William Thomson,[31] in Glasgow. This they set up in the little laboratory in Clyde Street and detected electric currents in isolated eyeballs when the light shone on them. The experiments required darkness and stillness, so Dewar went first to McKendrick's home each evening at around 10 o'clock, and then on to the laboratory where they worked until two or three in morning. This work led to three papers read before the Royal Society of

Edinburgh in April, May and June of 1873. The experiments went on until 1875 when colleagues often came to see these curious performances.[32] They included, on one occasion, Thomas Henry Huxley,[33] who was taking the natural history class at the university that session whilst Wyville Thomson was away in charge of the *Challenger* oceanographic expedition. McKendrick was later to say that Dewar's habitual sleeplessness was a factor that contributed to his ill-health in later life.[34] The principal result they found was that the potential between the retina and the optic nerve depended approximately logarithmically on the intensity of the light, an example of what is now called the Weber-Fechner law for the relation between stimuli and perceived effects. This work formed the subject matter of two Friday Evening Discourses at the Royal Institution in London that Dewar gave on 5 February 1875 and 31 March 1876.[35] They had extended their collaboration with studies on the physiological effects of various chemical compounds, including ozone and Dewar's favourite organic bases.[36] The gas, as they found by experiments on themselves, was damaging to living tissue. In taking up this line of research they may have again been following the example of Crum Brown.

Thomas Graham, when Master of the Mint, had found that palladium could absorb almost a thousand times its volume of hydrogen at atmospheric pressure and had concluded that the system so formed was an alloy of palladium and a metallic form of hydrogen that he called 'hydrogenium'.[37] Dewar had been interested in this 'alloy' since 1869 and in 1873 he measured its heat capacity. To do so he built a calorimeter which he enclosed in a brass vacuum jacket for its better insulation.[38] The paper is worth notice since the jacket was a forerunner of his much later and more efficient silvered vacuum flask that was to play a crucial role in his work on the liquefaction of gases.

Dewar and Tait showed that heated charcoal was, when it cooled, an efficient agent for mopping up the last traces of gas in a 'vacuum' vessel.[39] This property he and Dewar used in an experiment to study the spectacular 'radiometer' invented by William Crookes.[40] The discovery of the element thallium was Crookes's principal claim to fame and it was in 1871, when he was determining its atomic weight, that he noticed that his balance was disturbed by differences in temperature of the sample and the balancing weights, although the weighing was carried out in a vacuum. He followed up this anomaly by sensitive experiments with pith balls and so was led to the invention of what came to be called a radiometer. This was, and indeed still is, since it is known to us a scientific toy, a set of four vertical vanes mounted at the ends two horizontal and mutually perpendicular bars supported at the crossing-point of the bars on a pin about which the bars and their vanes can rotate freely about a vertical axis. One side of each vane was blackened and the other polished, with the black sides all facing in the same rotational direction. When light was shone on this device and when the vessel in which it was mounted was evacuated then it turned in the direction that it would have followed if the light were exerting a pressure on the blackened sides of the vanes.[41] At first Crookes thought, from his experiments with pith balls, that he had found some interaction between temperature and gravitation, but with the rotation of the vanes he inclined more to the idea that light exerted a pressure.[42] Whatever the explanation, he was convinced that he had uncovered some new physical effect. Such a claim, and his well-known interest in spiritualism, led George Stokes, the cautious Secretary at the Royal Society who managed its physical publications, to handle Crookes's papers with care, but eventually they were admitted to the *Proceedings* and, at last, to the more prestigious *Transactions*.[42] Maxwell had deduced that light should exert a pressure in his 1873 treatise on electricity and magnetism,[43] but this work was not known to the mathematically ill-trained Crookes, and, moreover, this effect would have led to the repulsion being greater at the white or reflecting surface. This was an effect that could not have been detected in the best vacua of the 1870s. Osborne Reynolds, the Manchester engineer, thought that the residual gas was responsible and that Crookes was doing no more than confirming the by-then generally accepted kinetic theory of gases.[44] His explanation was not easy to follow so Dewar and Tait entered the field. Their experiments, with their 'charcoal' vacua, confirmed those of Crookes but they gave a clearer account of how they thought the kinetic theory should be applied. The black surface was warmed by the incoming light and so gas molecules bouncing off its surface had a higher speed than most of those in the surrounding gas. At atmospheric pressure they hit one of these close to the surface and were returned to it, so leading to no net effect. But if the pressure were low enough the mean free path of the reflected hot molecules could be as long as the diameter of the enclosing vessel and so many of these molecules would proceed unhindered to the walls where they could lose their excess energy without affecting the vanes. The blackened surface thus received a net impetus. This paper was given before the Royal Society of Edinburgh on 12 July 1875, but was never formally published. It was Dewar who delivered the paper but the use of charcoal and the recourse to the kinetic theory of gases for the explanation of the effect both smack more of Tait than of Dewar. There was a long summary of the paper in *Nature*[39] which gave what seems now to be an adequate account of their conclusions, but with which Kelvin later expressed dissatisfaction.[39] Tait wrote to Norman Lockyer,[45] the editor of *Nature*, on the same day as the paper was read in Edinburgh, to say that: 'Dewar is to send you, by this post, an account of his and my paper drawn up by Scott-Lang, my assistant. He will put it right, and you will take care to head it *From a Correspondent*. Otherwise it might be looked upon as an infringement of the Royal Society of Edinburgh.' Two years later Tait wrote to Maxwell, as follows: 'As to Dewar & Tait. Their joint paper, which I still believe to be not only the *first correct*, but the *only correct*, explanation of Crookes &c. was prevented from appearing before the public in proper guise by those foes to knowledge, the Secretaries R.S., to wit Stokes & Huxley! True, I assure thee it was as thus – I was busy with an eloge of Στοξ [i.e. Stokes] for *Nature*, so Dewar undertook to give the paper, w^{h} he did uncommonly well.' Tait also expressed dissatisfaction with the article in *Nature* and it seems as if he and Dewar had tried to have the work published by the Royal Society of London but that it was turned down by the Secretaries, perhaps because the content had already appeared in *Nature*, as Tait had feared might have been also the reaction of the Royal Society of Edinburgh.[46]

The matter is more complicated than Tait and Dewar realised. A surface heated to produce a temperature gradient along it induces a flow of gas from the colder to hotter region, a phenomenon known as 'thermal creep'. In a radiometer the gas creeps from the outer edges of the black side of the vane towards the centre where it piles up since its viscosity prevents its instantaneous dispersion. The resulting small increase in pressure drives the vane back. Reynolds, and in more detail, Maxwell invoked 'thermal creep' and the flow of the gas from the edges of the black vanes in 1879, in one of Maxwell's last papers before his death later that year, but to this day a complete theory and its quantitative confirmation seem still to be lacking.[47]

One difficult experiment of Dewar's Edinburgh years was seriously in error. This work was done with William Dittmar,[48] an inorganic and analytical chemist for whom Dewar had a high regard, reinforced by the great emphasis both placed on facts and their distrust of 'speculation'. He was German-born and trained, but came to Britain and spent two short periods in Edinburgh during the second of which, 1872–3, he collaborated with Dewar. He moved briefly to Manchester and then, in 1874, to the chair of chemistry at Anderson's College in Glasgow. There he became known for his analyses of the sea-water samples collected during the *Challenger* expedition. He and Dewar tried to determine the density of potassium vapour by boiling it in an iron vessel at atmospheric pressure, sealing the vessel and then opening it in water and measuring the volume of hydrogen produced.[49] They concluded that the density required a diatomic constitution, K_2. In this they were mistaken, but in a later experiment at the Royal Institution, with his then assistant, Alexander Scott, Dewar found the correct monatomic constitution, K. In 1876 Victor Meyer had introduced his now well-known method of determining the density of the vapours of volatile liquids by using the vapour of a known mass of liquid to expel an equivalent volume of air from the apparatus. In 1879 Dewar and Scott adapted this apparatus to sodium and potassium, replacing their previous iron vessel with a more reliably inert platinum vessel, and so showed that both sodium and potassium had monatomic vapours.[50] Five years later, however, Henry Roscoe was saying that the spectroscopic study that he and Arthur Schuster had carried out at Manchester confirmed Dewar and Dittmar's conclusion that the vapour was diatomic![51] Dewar's Edinburgh years were marked by a volley of ingenious experiments over a wide range of chemistry but he never followed any of them up so as to develop a coherent line of research. Their versatility and sheer volume did, however, mark him out as an up-and-coming young man.

A third post had come Dewar's way in 1873. Since 1848 the Highland and Agricultural Society of Scotland had had as its chemist Thomas Anderson,[52] the Professor of Chemistry at Glasgow. He was the discoverer of pyridine and several pyridine bases, including picoline (Dewar's favourite molecule at that time), and was the author in 1860 of *Elements of Agricultural Chemistry*. But his health was now poor and on 8 January 1873 the Directors resolved to employ an assistant chemist at a salary of £100 to £150 a year. Dewar was appointed on 5 March.[53] His accession strengthened the position of the Society on the east coast of Scotland. On 21 January 1874 Anderson resigned and commended Dewar to the Directors as his successor. One of the duties of the chemist was the analysis of commercial manures, which were subject to adulteration by unscrupulous suppliers.[54] A year later Dewar reported that the number of analyses of manure, guano, animal feed and other substances had reached record numbers. The farmers were beginning to appreciate the efficiency of the new chemist. He extended his brief to cover drinking water and found many cases of farms where the supply was unsafe. In 1870, his brother Alexander, by then the Medical Officer at Melrose, sent him a sample of well-water that James said was a chalybeate water that 'was certain to recommend itself to the medical profesion'.[55] James held the post with the Highland and Agricultural Society until his move to Cambridge in 1875 but he had by then become disillusioned by the low intellectual level of the analytical work he was asked to do. In his letter of resignation he wrote: 'I still trust, however, that the Society will ultimately see that that this office of Chemist will never be properly filled except by one thoroughly trained in scientific research, and this, the making him a real agricultural chemist will depend on the means placed at his disposal for applying his scientific knowledge to agriculture.'[56]

By April 1874 Anderson had vacated the Glasgow chair also, and the editor of *Nature* hoped that the Home Secretary, whose responsibility it was to fill this Regius chair, would choose someone who would carry out original research.[57] Dewar at once seized the chance of applying for one of the most important chairs of chemistry in the country. The procedure was then for a candidate to write to as many colleagues and acquaintances as he could summon and ask them to send him a reference. He then printed these as a pamphlet and sent it to the electors. It was not an efficient way of choosing the best candidate since the open references, or testimonials, were usually bland and uninformative. Faraday politely turned candidates away and generally refused to give references. But Dewar went about the job of collecting them with his usual efficiency. The first, from Kekulé, is dated 13 April 1874, and he had a list of 17 when, four days later, he sent his application to the Home Secretary. Two supplementary series of references followed and by 15 May he had submitted no less than 53; most laid great stress on his ability as a lecturer. He had also by then 28 publications to his name and he emphasised in his letter of application that 'I have not confined my researches to one department of Chemistry, but believing a Professor of that science in connection with a medical school ought be conversant with its application to Physiology, I have made several contributions with a view to the progress of Medical Science in its relation to Physics and Chemistry'.[58] His numerous papers did indeed cover an extraordinary wide range from physics to physiology but they lacked any coordinating viewpoint. It seems as if he were concerned to establish facts but not to draw any but proximate conclusions from them. The question 'Now what are the facts?' was a phrase he was fond of using in his lectures and papers. He was, however, by then a man with a growing reputation who was on the Council of the

Royal Society of Edinburgh and, soon, also of the Chemical Society of London. The range of testimonials is evidence of the wide impression his work had made. But all this trouble was apparently in vain; the candidate chosen in August 1874 was John Ferguson, a man five years Dewar's senior, who had been Anderson's assistant at Glasgow. He is known today as a historian and a bibliographer of chemistry, the author of *Bibliotheca chemica* (1906), a catalogue of the alchemical and chemical library of James Young. He was not to contribute much in the way of original chemical research during his long tenure of the Glasgow chair. But Dewar's industry in compiling references had not been entirely in vain for early in 1875 the ailing Robert Willis died in Cambridge, thus leaving a vacancy in the Jacksonian Professorship of Natural Experimental Philosophy.

Chapter 3
Cambridge

The Rev. Richard Jackson of Torrington, Herefordshire, died in 1782 and established in his will a professorship at Cambridge which was hedged around with numerous provisions, the principal one of which was that it should be devoted to 'Anatomy, Animal Economy, Chymistry, Botany, Agriculture or Materia Medica' and that the holder should 'have an eye more particularly to that *opprobrium medicum* called the gout'.[1] Over the years these conditions had become ignored and in 1874 the holder, Robert Willis,[2] was an authority on 'mechanism' or applied mechanics and physics. The formal title of the holder of the chair was the Jacksonian Professor of Natural Experimental Philosophy. Willis had been ill for some time and a deputy, Rev. J.C. Williams Ellis, had been appointed for the last two years. When Willis died on 28 February 1875 the University was divided on the field the next holder should profess. Some thought that the status quo should be accepted and that a branch of applied physics should be maintained; others pointed out that more regard should be accorded to the conditions of the will. The Council of Senate[3] decided that:

> the practice of the Professorship should be brought into closer conformity than hitherto with the general objects indicated in the terms of the foundation. They are moreover of the opinion that will be advantageous to the University, as well as in direct conformity with the design of the Professorship, that the lectures of the Professor should be directed, at least in part, to the illustration and advancement of some branch or branches of Applied Physics.[4]

This self-contradictory report was interpreted by *Nature* as a call for the next holder of the chair to be in a field similar to that of the late Professor Willis.[5] This view was supported in a paper, dated 8 March, circulated by Alexander Freeman, St John's College, who said that Jackson's conditions were now 'fanciful and obsolete'.[6] There was, however, a powerful lobby calling for a chemist and preferably with one with the qualifications to teach this subject to the medical students. This lobby was headed by George Downing Liveing,[7] the Professor of Chemistry, and George Humphry,[8] the Professor of Anatomy. They attracted the influential support of George Gabriel Stokes,[9] the Lucasian Professor of Mathematics, who, as Tait put it when Dewar's name came forward, 'took it into his head to start him for the Jacksonian, and kept us all alive with telegrams'.[10]

The Vice-Chancellor gave notice that the election was to be held on 13 April, the electors being the members of the Regent House, that is, essentially the 300 resident MAs who comprised the teaching staff of the University.[3] The issue was debated at a meeting of the Senate on 13 March.[11] Peter Latham, the newly elected Downing Professor of Medicine, wanted a Professor of Public Health or Preventive Medicine, Freeman repeated his call for a physicist, Humphry wanted an organic chemist or a physiologist, and Henry Cookson, the Master of Peterhouse, wanted a chemist, saying that 'Mechanism could not be found in the will and mechanism was not research'. Liveing, in a paper written from Weymouth, naturally wanted another chemist, a plea he had been making for some time.[12] Stokes added to the confusion by saying that in Jackson's time chemistry was part of physics.[13]

Both Cookson and Humphry deplored the mode of election by such an unqualified body of men as the 300 or so resident MAs, and there was criticism of the rule that a vote must be held within 28 days of the Vice-Chancellor giving notice of the vacancy, but it was made clear that neither condition could be varied without new legislation. In the 1850s two Royal Commissions, and the consequent Executive Commissions, had criticised Oxford and Cambridge for this democratic but hasty and uninformed way of choosing professors. (When Maxwell died on 5 November 1879 there was a petition circulated 10 days later, which Dewar signed, asking that Rayleigh be chosen as the next Cavendish Professor; he was elected on 12 December.[14]) Oxford had abandoned these primitive rules at the election of William Odling as Professor of Chemistry in 1872 but Cambridge did not opt for a qualified electoral board until 1883, two years after Liveing had been made an external member of the Oxford board. Charles Heycock, a colleague of Liveing, was to say many years later that 'there was a good deal of wire-pulling' over Dewar's election at Cambridge.[15]

With Willis, the chair had been under the Mathematical Sciences Board but it was now transferred to the newly created Natural Sciences Board, and the holder was required to live in Cambridge for at least 18 weeks each year, to lecture in two of the three terms, and to give not less than 40 lectures a year. The chemistry faction gained the upper hand and their attention was drawn to Dewar, possibly because of the publicity he had received when compiling his list of 53 referees to support his application for the Glasgow chair the previous year. He was to use this list again at Cambridge, noting however that he had now published 30 papers. Moreover he had also been honoured by the Chemical Society with the invitation to give a lecture, which he had recently done on 19 March, choosing as his subject the fashionable one of 'Dissociation'. That year he was also elected a member of Council of the Society.

Stokes wrote to him to sound him out and Dewar replied on 6 April, thanking Stokes for considering him for the post and for his advice.[16] It seems that Humphry then went to Edinburgh. Hamilton Dickson,[17] who was to become Dewar's brother-in-law, recounted many years later a dramatic story that Helen Dewar and James's brother, Alexander, were in the post office one day when, by chance, they saw the draft of part of an out-going telegram: 'Found the man you want here, Dewar, telegraph at once. Humphry.' Dickson says that James, knowing nothing of Humphry's message, received at lunch a telegram offering him nomination for the post and asking for an immediate reply since the electors were meeting that afternoon.[18] This cannot have been the formal meeting of the electors to be held on 13 April but was probably the informal meeting in the Guildhall of those supporting the election of a chemist with medical interests that Liveing had convened for the afternoon of 6 April. This meeting did not finish its business that day, possibly because they did not hear from Dewar in time; his letter to Stokes could not have arrived until the next day. An adjourned meeting on 7 April, chaired by Cookson, passed a motion moved by Stokes and Liveing: 'Having regard to the requirements of Mr. Jackson's will as re-enacted in the Statute of 1861, and to the recent decision of the Senate by which the Jacksonian Professor is made *ex officio* a member of the Board of Natural Science Studies, this meeting recommends as a fit and proper person to be elected to the Jacksonian Professorship Mr James Dewar, Demonstrator in Chemistry in the University of Edinburgh, and formerly assistant to Professors Tait and Lyon Playfair.'[19]

By then five other candidates had come forward: Henry Armstrong (1848–1937),[20] J.C. Williams Ellis (1833–1913),[21] Walter N. Hartley (1847–1913),[22] Edmund J. Mills (1840–1921)[23] and James Stuart (1843–1913).[24] Armstrong, then the Professor of Chemistry at the London Institute, was the youngest and might have been the best choice, although he was to develop heterodox views in later years. He had support from some eminent chemists – Frankland, Roscoe, Brodie, Debus and Kolbe, and had initially been Liveing's first choice but, as he told Armstrong, he felt obliged to follow his medical colleagues when they preferred Dewar.[25] Hartley was to make his name later in the century as a spectroscopist but was also young and held only the post of Demonstrator at King's College, London. Mills, an Assistant Examiner at the University of London, was later Professor of Technical Chemistry at Anderson's College and a founder member of the Society of Chemical Industry in 1881. He had a strong reputation but was perhaps too 'technical' for Cambridge and seems never to have aroused the enthusiasm of the active electors. Ellis and Stuart were former high Wranglers but both came from the mathematics and physics background that the dominant group of electors had turned down. When the Guildhall meeting came out so decisively in favour of Dewar the other candidates withdrew and the actual election on 13 April was such a formality that he came down from Edinburgh for the occasion. This was described in the following words:

> The candidates were H.E. Armstrong, Prof[r] of Chemistry in the London Institute, J. Dewar, Demonstrator of Chemistry in the Uni[v] of Edinburgh, J.C.W. Ellis, M.A., Sidney, 3rd Wrangler in 1856, W.N. Hartley, Demonstrator in Chemistry in King's College, London, E.J. Mills, J. Stuart, M.A., bracketed 3rd Wrangler in 1865.
>
> All retired before the election on 13th of April, 1875, excepting Dewar, who was therefore elected without opposition, 11 votes being given for him, none for anyone else.
>
> Immediately after the election on 13th of April, 1875, M[r] James Dewar appeared in the Senate House, made the declaration of office, and was admitted on his knees to the Professorship by the Vice-Chancellor.
>
> This was done in my presence,
>
> H.R. Luard
> Registrary[26]

This admission ceremony might have been a trial for the slightly lame Dewar and it seems that he was the last professor to be admitted in this antiquated fashion. The 32-year-old professor was still without a university degree but Cambridge at once conferred a MA on him and fixed his salary at £500, the same as that of Liveing. Only £140 of this came unconditionally from Jackson's endowment, the rest was dependent on his conforming to the conditions laid down for residence and lectures. The appointment was welcomed by *Nature* in terms flattering both to Dewar and to the University.[27] Lyon Playfair wrote the next day to congratulate him, adding 'I hope that Glasgow University will now feel that a mistake was made in not appointing you'.[28]

The University decided at once not to abandon the field that Willis had professed, and it may have been the knowledge that moves for a replacement were underfoot that led to the precipitate retirement of some of the candidates for the Jacksonian chair. A Grace of the Senate of 25 May established a new chair of Mechanism and Applied Mechanics for which the election was held on 17 November. This time James Stuart beat the mathematical coach E.J. Routh by 111 votes to 81, with Willis's former deputy, the unfortunate Williams Ellis, coming last with 36 votes.[29]

When Dewar arrived in Cambridge the University was just beginning to take experimental science seriously. William Whewell, the Master of Trinity College until his death in 1861, had laid down that the fundamental basis of a university education was classics and mathematics, pure and applied. These were deductive subjects whose content was, he supposed, fixed. The experimental sciences were 'progressive', since their content and even their foundations were still changing, but they showed how the world was constructed and should be studied once the foundation subjects had been mastered. In 1848 the Natural Sciences Tripos was established as an honours degree that could be taken after the Mathematical Tripos. Six candidates sat the examination in its first year in 1851. The highest marks were obtained by Liveing who realised then that he would also need some laboratory experience if he were to take chemistry seriously. After a year's work in London and Berlin he returned to Cambridge, serving from 1859 as deputy to the now ailing Professor of Chemistry, James Cumming, and succeeding him as Professor on his death in 1861. In the same year the Natural Sciences Tripos was recognised as a first degree but Liveing was still dissatisfied with the lack of provision for laboratory instruction. He was to write later 'The University possessed no laboratories, and provided no appliances for her professors, much less for students, and did not in all cases even provide lecture rooms'.[30] On his return he had held a class in practical chemistry for medical students in a primitive laboratory in Slaughterhouse Lane (now Corn Exchange Street). This he had set up at his own expense, but St John's College, of which he was now a Fellow, came to his aid and built for him a better laboratory behind their New Court. Meanwhile the Jacksonian Professor had been provided with a small room, badly lit and without a fireplace, and a shared lecture theatre, in a building of 1784 on the site of the former Botanic Gardens, in the corner bounded by Downing Street and Corn Exchange Street, a spot now opposite the Museum of Archaeology and Anthropology. In 1833 five small rooms for the Chemistry and Jacksonian Professors were added on the north side of this building and some further provision for the medical staff was added to the south. By 1853 it was clear that a more radical solution was needed. A syndicate, of whom Willis was the most active, was appointed to report. On 31 December he set out in detail what he saw as the needs of the different professors, noting first that 'The University at present possesses no apparatus for physical or mechanical science, with the exception of that employed in the Observatory'. The needs of the Chemistry and Jacksonian Professors were described, as follows:

> The Chemical Professor requires a laboratory and private room with a Balance Room, and the other usual appendages. The modern method of teaching Chemistry is to instruct the Students in the actual practice of the laboratory, in addition to the course of lectures. For this purpose a Students Laboratory must be constructed, separated from the Laboratories and Lecture-Room of the Professor, and fitted with distinct working places for the Students. A room about forty feet long, and fifteen wide, will accommodate fourteen Students, but the plan should comprise the easy addition of more room if required. ... The apparatus required by the Jacksonian Professor for lectures on manufacturing machinery, steam engines, &c. is necessarily bulky. Much of it may be disposed in the Museum of Apparatus, but he requires, in addition, a room of the nature of a workshop, in which apparatus may be fitted together for use in the Lecture-Room; and the frames and unsightly portions of lecturing machinery stowed away. This room should be contiguous to the Lecture-Room. He also requires a private room with a small laboratory fitted with a forge, and a furnace for melting metals, &c.[30]

This ideal situation did not come about in 1853; there was no money. The description is, however, worth note as a statement of how Willis saw the problems at that time. He accepted that students of chemistry need 'hands-on' experience in the laboratory although he seemed not to wish to provide this for his own students. The apparatus for both subjects seems to be for teaching; there is no specific mention of research. The needs of the Jacksonian Professor were tied to Willis's own interpretation of the field of study, although they could well serve as a description of what Dewar was eventually to acquire for himself at the Royal Institution in the 1880s.

In 1865 a two-storey building was put up on the site but this did not provide space for the chemists although they may have benefited indirectly from the reshuffle that then took place. More useful was the erection in 1872 of a students' chemical laboratory over the Jacksonian Professor's rooms, but there was still no adequate provision for experimental work by Dewar. It was noted then that he shared with the Professor of Chemistry 'the use of chemical lecture-room and the rooms adjoining', whilst Stuart, the new Professor of Mechanism and Applied Mechanics, used some of rooms previously assigned to the Jacksonian Professor. Nevertheless it was 1878 before Stuart was able to introduce practical classes for his students. Several colleges had followed St John's in providing laboratories for their chemistry students – Caius[31] in 1873 and Sidney Sussex, Girton and Newnham later in the decade. But none of these laboratories was open to the Jacksonian Professor nor to his assistant, whose appointment the Museums and Lecture-Room Syndicate had previously authorised.[32]

The overcrowding and lack of equipment in Cambridge contrasted with the more extensive, if no better funded, facilities to which Dewar had become accustomed in Edinburgh and was the start of his dissatisfaction with Cambridge. Henry Armstrong, his closest scientific friend in his later life, thought, moreover, that Cambridge and Dewar were temperamentally unsuited to each other; the sober university did not take kindly to the free-and-easy ways of the young Scot. (It is hard to see in this description of Dewar the formal and dignified the professor that he soon became but Armstrong knew him well and must have had good grounds for his characterisation.) Nevertheless the appointment had been made and Dewar and his wife moved south. They lived for a time at 19 Brookside, a street to the east of Trumpington Road, but moved, probably in 1885, a little further in towards the centre, maybe renting at first but eventually buying, from Caius College, 1 Scroope Terrace, in a row of handsome early nineteenth-century houses on the west side of Trumpington Street.[33] The Terrace was a popular location for the academic staff; the Maxwells[34] lived at No. 11 and later J.J. Thomson[35] and his family took No. 6. The Dewars' former home now houses the Department of the History of Art.

In the autumn of 1875 Dewar settled down to his courses of lectures. The academic year was divided into three terms: Michaelmas Term, from October to December, Lent Term, from January to March, and Easter Term, from April to June. He chose at first to lecture only in Michaelmas and Lent Terms, on Tuesdays, Thursdays and Saturdays. In his first year the opening course was on 'Dissociation and Thermal Chemistry', and the second course, on 'Organic and Animal Chemistry'.[36] 'Dissociation' is today a term that carries little theoretical depth; it merely describes the breaking up of a molecule into simpler molecules, usually by heat. In the 1870s, however, in the primitive early days of chemical thermodynamics, it encompassed ideas that we now group under such names as the law of mass action, Le Chatelier's principle and the second law of thermodynamics itself. The founding book of the subject was Henri Sainte-Claire Deville's *Recherches sur la Décomposition des Corps par la Chaleur et la Dissociation* of 1860, a book much admired by Dewar. He lectured on the subject of dissociation before the Chemical Society in 1874 and 1882, before the British Association in 1878, and to the Royal Institution with course of five lectures in 1879 and a Friday Evening Discourse on Deville on 21 April 1882.[37] It was also the subject of his early undergraduate lectures at Cambridge. These contributions describe mainly the changes of phase that accompany the dissociation of solids by changes of temperature and of the pressure of the surrounding gas. The only theoretical aspect of the subject that he considered was the use of Clapeyron's equation and a passing mention of Willard Gibbs in 1882.

From the start of the academic year of 1876–7 he alternated between a course on physical chemistry in Michaelmas Term and one on organic chemistry in Lent Term, but the days of the week changed from year to year since he now had to time his days in Cambridge around the duties he had taken on by his additional appointment in 1877 to the Fullerian chair of chemistry at the Royal Institution. His two courses were advertised in the *Cambridge University Calendar* for 1877 with the words: 'Students attending both courses have an opportunity of hearing lectures on the whole range of chemical facts.' These words were surely written by Dewar himself. His whole career was to be built around determining the facts; theory he seemed to think of as little better than speculation. Moreover he might have wished to distinguish what he wanted to teach the students from what Liveing, the Professor of Chemistry, taught. Liveing's principal course of lectures in the year that Dewar was appointed was entitled 'Principles of Chemistry'.[38] Dewar was to make his name in London as a fine lecturer to both scientifically trained and to inexpert audiences, but he was less successful in Cambridge. The undergraduates wanted clear lines of reasoning that they could transmit into their notes, and their appetite for undigested facts may not have matched his. 'Not a great teacher' and 'often an incoherent lecturer' was Henry Armstrong's verdict. In University politics he generally sided with the reformers, being, with Liveing, one of 13 of the 34 University Professors who called for a reform of the Statutes, and four years later, with both Liveing and Rayleigh, signing a petition for the admission of women to examinations and degrees.[39]

It may have been a common interest in the factual side of physics that led Frederick Guthrie[40] to propose Dewar for election to the Royal Society in December 1876. Guthrie, who had trained as a chemist, was Professor of Physics at the Government-sponsored School of Mines in Jermyn Street. He had been the principal mover behind the foundation of the Physical Society in 1874, after he had been dissatisfied with the reception of his work by the Royal Society. This new society attracted many schoolmasters to its membership and its meetings were noted for their emphasis on experimental papers; Clerk Maxwell declined to become a member. Guthrie was joined in proposing Dewar for the Royal Society by 23 other Fellows who signed the certificate; these included Thomas Andrews, Frederick Abel, William Crookes, Clerk Maxwell, William Thomson, John Tyndall, Lyon Playfair, and some who will appear later in Dewar's life. Two obvious names that do not appear are those of his main supporters for election to the Cambridge chair, George Liveing and George Stokes.[41] Liveing's absence is not easily explained but, by custom, Stokes was unable to sign since he was a Secretary of the Society. Nevertheless Dewar's backing was impressive and he was elected at the first opportunity, on 7 June 1877, when there were 57 candidates and 15 elections.

Chapter 4
Demonstrators

Cambridge recognised that professors of the experimental sciences needed help in preparing lecture demonstrations, even when there was no provision for 'hands-on' experimental work by the students themselves. This need led to a report to the Senate from the Museums and Lecture-Rooms Syndicate of 11 November 1875 in which they proposed an assistant at £100 a year for the newly appointed Jacksonian Professor.[1] They recommended that 'such assistant to be at all times under the order of the Jacksonian Professor, and his appointment and removal to be made in accordance with the powers of the Syndicate'. This recommendation was accepted and written into the Ordinances, with the proviso 'that the Syndicate have the power both of appointing and removing the assistants and servants of the museums and lecture-rooms, but not without the concurrence of a majority of the whole Syndicate'. In an account in his *Pamphlet*[1] of the subsequent row over the employment of his second assistant Dewar attached a footnote to this phrase noting that the University Calendar had a different account of the conditions, saying that 'the appointment is made by the Professor with the consent of the Vice-Chancellor, and the person so appointed is removable in like manner'. These contradictions were to cause problems later.

Dewar's first choice as his assistant, Alexander Scott[2] from Edinburgh, stayed with him for nine years and the two seemed to have worked amicably together throughout that time. Scott's career was to be closely bound up with Dewar's. While still a student, he had been Dewar's assistant in his post at the Veterinary College in Edinburgh. He worked as Dewar's assistant in Cambridge from 1875 to 1884, when he went to Durham and Middlesborough as a school master and where he taught in two schools simultaneously. He had graduated while at Cambridge, obtaining first-class honours in Part 1 of the Natural Sciences Tripos. He returned to Cambridge in 1891 as Dewar's Jacksonian Demonstrator and later moved to London to join Dewar at the Royal Institution. He was proposed for election to the Royal Society in 1894, and elected in 1898.[3] (Dewar did not sign his certificate.) He was a careful research worker who specialised in the determination of atomic weights, then often a controversial matter when their closeness to integers, and their numerical ordering and its bearing on Mendeleev's Periodic Table, were still matters of debate. His determination, for example, of the atomic weight of tellurium at 127.6 placed it firmly above that of iodine at 126.9, although the latter's membership of the halogen group required it to be heavier than tellurium if Mendeleev was to be believed.[4]

Scott's departure for Durham in 1884 left Dewar in a quandary; in what field should he look for a successor? Dewar's lectures, like his research at the time, covered a wide range of organic, inorganic and physical chemistry but he may have been conscious of the way in which organic chemistry was coming to dominate the field, particularly in Germany. Physical chemistry was growing but was still seen as a new subject. Whatever his reasoning, he decided that, after Scott's departure, his next assistant should be an organic chemist and turned for advice to the accepted leader of that field, August Wilhelm von Hofmann[5] in Berlin. The man whom Hofmann recommended was his own assistant, Siegfried Ruhemann,[6] who had recently received his Ph.D. after working under Hofmann's supervision. Ruhemann was Jewish and might have had difficulty in obtaining a permanent position in Germany at that time, moreover he was unmarried and free to move. He came to Cambridge in 1885 and almost at once took over from Dewar the lectures on organic chemistry although lecturing was not formally part of his duties. He lectured at first three times a week, but soon increased that load until he was lecturing nine times a week, in fractured but quite intelligible English, and was running a practical course also. His formal duties did not require him to undertake his own programme of research, but again this he did. He had been given an Hon. MA[7] and had become a member of Caius, which had had its own laboratory since 1873 in which Ruhemann was to work. It was to prove a fortunate choice of college. From 1888 the whole of chemistry benefited from the erection of a new departmental building in Pembroke Street in which Ruhemann had rooms on the top floor. From his arrival in 1885 until his dismissal six years later he published 13 papers, either on his own or with students. In spite of these achievements Dewar regarded him solely as an assistant without any rights of his own – Ruhemann, he believed, held his appointment solely at Dewar's pleasure. This was a view of an assistant's position that he had long held, as was clear from the sequence of events after Maxwell's death on 5 November 1879 and the election of Rayleigh a few weeks later. Rayleigh had never held a university post before and wrote to professorial colleagues to ask for advice. Dewar's reply was that he should ensure that his assistants carried out his instructions and should not be allowed to embark on research ventures of their own. According to Arthur Schuster, Rayleigh was appalled by this reply.[8] Dewar was not alone in his attitude to the work of his assistants; when Crookes wanted to publish a translation of a German work on hygiene he asked his assistant, James Gardiner, to undertake it. Gardiner was a man with a London MA and was a Fellow of the Chemical Society, but Crookes published the work under his own name only.[9]

Dewar and Ruhemann fell out at an interview held on 1 December 1890, and Dewar wrote to him on 7 December to say 'that having regard to all the interests involved I can see no other way out of the difficulty. I shall therefore be obliged by your sending me a formal resignation of the office you now hold'. Ruhemann replied the next day to say 'that I mentioned during our interview that I should not give up the position I hold until I had got a suitable appointment. I am sorry that I cannot as yet agree with your request'. Dewar's response was to say that he cannot 'allow a condition of affairs eminently unsatisfactory to the Chemistry Department to continue at your discretion until such time as you find suitable appointment … [To do so] would be a serious dereliction of duty on my part. Unless therefore I receive a formal resignation I shall be obliged to send a dismissal in accordance with the University Rules'. There was no mention of any specific cause of the rift and Dewar avoided such mention throughout the affair, but he is clearly implying here that it was a general problem that affected the whole of the Chemistry Department and not merely a personal disagreement between himself and Ruhemann. In doing this he would have had the support of Liveing, with whom he had already discussed the situation and who was to back him up throughout the quarrel. There is, however, no other evidence that the argument was anything but a personal matter between Dewar and Ruhemann. The root of this must have been in Dewar's treatment of Ruhemann as a servant and not as a colleague. There are, however, letters that show that a specific cause was the fact that Dewar required Ruhemann to carry out chemical tests and analyses for Dewar's considerable outside consultative business. Pattison Muir was in charge of the laboratory at Caius, and so in a position to understand Ruhemann's work for Dewar. He wrote to Scott, now in Durham, to get evidence on this point a week after Dewar's exchange with Ruhemann described above:

> Dear Scott, 13 December 1890
>
> You may have heard that Dewar and Ruhemann are at daggers drawn. R. has been complaining of many things especially of the commercial analyses that Dewar has given him to do. Dewar threatens to dismiss Ruhemann; Ruhemann refuses to resign. The matter will now come before the Museums & Lecture room Syndicate who alone can dismiss Ruhemann. I am very curious to know whether commercial work was done in Dewar's lab. in your time. Would you mind telling me?
>
> I think that we are in for a splendid row; many men devoutly wish it may end in Dewar's not Ruhemann's dismissal.
>
> I am, yours very sincerely, M.M. Pattison Muir[10]

The next day, Muir's colleague, Arthur Sheridan Lea, the Fellow in Physiology at Caius, wrote even more directly:

My dear Scott,

> I shall be greatly obliged if you will let me know whether [during] the time you were Dewar's assistant he employed you to carry out analyses for him of articles for purely commercial purposes. Can you give me an idea of (i) the kind of commercial analyses you performed and (ii) the extent to which your time was taken up with that work? We may shortly have to consider Dewar's action in using the University Laboratory & Appliances for purely commercial work, and I therefore wish to have some information as to how things were going on here in your time.[11]

We do not have Scott's replies but it may be significant that he kept these two letters for over 20 years until a time when he too was to be embroiled in a similar dispute with Dewar. There is also a list, compiled by John Foster, of the letters from Dewar to Ruhemann from 1886 to 1890 on analyses of water, wines and dyestuffs, and of work on incandescent lamps.[12] The evidence seems conclusive that it was Ruhemann's exploitation as an assistant to Dewar's commercial practice that was a specific, and maybe *the* specific cause of his dissatisfaction, and that Dewar was at first loath to reveal to the University how much of Ruhemann's time this was taking, or to discuss whether it was a proper use of University facilities. Later he was more forthcoming in his correspondence, although never in discussions with the various university committees that became entangled in the dispute.

Dewar wrote to the Vice-Chancellor, the Rev. Montague Butler, Master of Trinity College, on 15 December to tell him of the 'strained relations' and to ask him for 'the authority of the Museums and Lecture-rooms Syndicate to appoint someone else in his place'. He offered to pay out of his own pocket for a quarter's salary to Ruhemann in lieu of notice. The Vice-Chancellor replied with two letters, an official one asking for a postponement of any action until the next term, and a private one saying that he had learnt in the last 10 days 'that not a few very distinguished members of the University have the strongest sense of the value of Mr. Ruhemann's services as a teacher, and that they would regard his retirement from Cambridge, especially if it seemed to carry with it anything of a stigma, as little short of a calamity. He is beloved and respected in no ordinary degree, and his dismissal would awaken very general and earnest sympathy'. These were strong words from a Vice-Chancellor and cannot have pleased Dewar. In reply, on 22 December, he assured the Vice-Chancellor that 'the removal of Mr. Ruhemann will in no way cause the calamity to the teaching of chemistry in the University prognosticated by the distinguished members of the University to whom you refer'. At a later meeting of the Senate, Sedley Taylor of Trinity College, a musicologist, produced figures to show how popular Ruhemann's lectures had become. In the session 1890–1 Ruhemann had classes of 87 and 91 in the Michaelmas and Lent Terms, while in the term just finished Ruhemann had 118 and Dewar had six in their classes. Liveing responded by saying that Dewar had had large classes in the past and would no doubt have them again. By 1890 Dewar's activities were centred on the Royal Institution in London and it may have been some resentment on his part of Ruhemann's success as a teacher in Cambridge that fuelled the rift between them. Meanwhile Ruhemann had been to see Liveing who wrote to Dewar to try to effect a resolution in which Ruhemann would obey Dewar's orders; 'short of being a servant, he would do anything you wished'. But Dewar was not satisfied with Ruhemann's apology and would not accept it.

The Syndicate met on 26 January 1891 and, not unnaturally, asked Dewar to tell them what was Ruhemann's offence. This he refused to do, maintaining that he was following the rules and that 'it is only the Professor who is entitled to define the duties of his assistant, and with him rests the decision as to whether the obligations have been properly discharged'. The Syndicate replied that they could not act 'without some clearer indication of the offence charged'. Dewar suggested that the Syndicate should confer with Liveing who knew all the circumstances, but Liveing was not prepared to go beyond what Dewar had said in public. By March the matter had reached a committee of the Senate appointed to try to resolve the situation. Its chairman, Austen Leigh, the Provost of King's College, asked Ruhemann if he had ever refused to follow any order from Dewar. He said that he had not, but that he had 'constantly protested against certain orders which entailed work in no way connected with the University, although he had executed such orders and received extra fees for them'. Dewar denied the accuracy of the statement about 'constant protests', saying that the fees had stopped after Ruhemann had complained to Liveing at Dewar's use of his services.

In the middle of all this argument the Chemical Society celebrated its jubilee on 25 March with an eight-course dinner at the Hotel Métropole in London. Both Dewar and Ruhemann attended, but fortunately were seated on different tables. Dewar was, in fact, placed directly opposite William Ramsay with whom he was later to have a controversy that became much more notorious than his dispute with Ruhemann. By April Dewar had heard of the letters that Muir and Lea (although he does not name them) had written to Scott and said that they should instead have written to him, an alternative that was hardly open to the two men who were gathering evidence to be used in Ruhemann's defence. He told the Provost of King's that he never did commercial work in Cambridge, although he said that he had advised courts of law. He does not say so, but it may have been in preparing to act as an expert witness that he had ordered tests or analyses to be made. He seems to be defining 'commercial' narrowly if he excluded payment as an expert witness and for the analyses associated with his court work. He lists the 16 payments made to Ruhemann in the last five years; they vary from three guineas [£3.15] to £12. Ruhemann's acceptance of the fees must have undermined the strength of his case, but in May the Special Board for Physics and Chemistry recommended that a university lectureship in organic chemistry be created for him, although they recognised that Liveing would not provide space for his lectures or research. Fortunately Caius was able to come to the rescue and provide these facilities. His salary was to remain unchanged at the £100 he had received as Dewar's assistant. The General Board of Studies agreed to this appointment in a report signed by the Vice-Chancellor and 11 other members, who included George Humphry and J.J. Thomson. On the same day, 13 June 1891, they suspended any re-filling of the now vacant assistantship. The whole correspondence and the reports occupy 58 pages of Dewar's *Pamphlet* which he issued in October. He seemed to think that he emerged with credit from this complicated affair for he sent a copy of his pamphlet to Thomson in Glasgow:

> 15th October 1891
>
> Dear Sir William Thomson,
>
> Accept my warmest thanks for the magnificent set of original papers you kindly forwarded this morning. Long may you live to create such noble works.
>
> I enclose for you a pamphlet dealing with a University Dispute that may interest you. It came about by the conduct of the late V.C. He behaved in the most extraordinary way to both Liveing and myself. The whole affair is most discreditable to the University. The place is infested with a clique of Log Rollers chiefly organised by "**Michael**" and his physiologists.
>
> Ever yours truly, James Dewar[13]

Dewar writes as one who has been wronged by the University and whose stance should be admired. Thomson, Tait and Dewar were all Fellows of Peterhouse, the last having been elected in 1883. The 'late V.C.' of his letter to Thomson was Montague Butler, who was still alive but had come to the end of his stint as Vice-Chancellor. 'Michael' was Michael Foster, the Professor of Physiology, and Secretary of the Royal Society, where he had succeeded his teacher Thomas Henry Huxley. He was then a man of radical views and so probably not to Dewar's taste, but he seemed to have played no particular part in the dispute with Ruhemann, nor to be related to J.E. Foster.[14]

Ruhemann continued as a lecturer in organic chemistry but was able only to take his place fully and openly in the University Chemical Laboratory in 1908, after Liveing's retirement. His career prospered, he married and had a son, Martin. Siegfried took British citizenship, and in 1914 was elected to the Royal Society. His proposer and seconder were Norman Collie and William Ramsay who were at University College, London, and by now sworn enemies of Dewar. The xenophobia that marked the outbreak of war that year soon made the Ruhemanns' professional and social position in Cambridge extremely uncomfortable; he resigned his lectureship, lived quietly, and returned to Germany in 1919, where, after a spell as Emil Fischer's assistant, he became head of an industrial laboratory. There is no evidence that Dewar himself was in any way involved in the harassment of the Ruhemanns in Cambridge in 1914–15. Unlike William Ramsay, he seems to have been relatively free of xenophobia, although he removed his name from lists of members of German scientific societies during the war. Ruhemann had retained his Fellowship of the Royal Society throughout the war but he allowed it to lapse in 1924 and he never received an obituary from the Society. Martin took a doctorate in physics in Berlin in 1928 and became an authority on cryogenics and gas liquefaction, which was just the field in which Dewar had established his reputation.[15] By 1938 the family was again persecuted in Germany, this time for their Jewish family relationships. So back they came to Britain, where Siegfried Ruhemann lived until his death in London in August 1943.

The 'affaire Ruhemann' damaged Dewar's reputation in Cambridge, although he never seemed to have understood this. Doubts about his conduct in a law case with Alfred Nobel (Chapter 7) can only have completed the disillusion in the years 1891–5. Dewar continued to fulfil his statutory duties by giving the required stint of lectures each year but his research and now most of his scientific life centred on London where he had been elected to the Fullerian chair of chemistry at the Royal Institution in 1877. He did however continue with miscellaneous lines of research at Cambridge and for these and for help with his lectures he continued to require an assistant. But conventions were changing, the title of the professorial assistants was altered to the more professional one of 'Demonstrator'[16] and James Porter, Master of Peterhouse after Cookson's death in 1876, brother-in-law of Guthrie Tait, and himself a mathematician, was able to declare at a meeting of the Senate that 'there was no doubt that the assistants originally belonged to a lower order than at present'.[17] So when Dewar's post was released he sought the return of Alexander Scott as the Jacksonian Demonstrator. Scott knew of the circumstances of Ruhemann's dismissal and approached the proposal gingerly. Dewar wrote to him on 8 August 1891: 'I see no use [of?] the enclosed agreement. Your duties in the future will in no way differ from what they were in the past. As to liberty between Terms and small things I make no [direction?]. All I am interested in is to maintain harmony of work and simple good feeling. You will see Liveing at the B[ritish] A[ssociation] and can discuss matters with him. It is clear that we cannot allow Caius College to get a monopoly of teaching Organic Chemistry.' A few days earlier he had written: 'With regard to Lectures it is the advanced Organic that you must take in hand. I have myself down for the Elementary course and I may take Thermal & Dissociation although the Physical side is [now?] done.'[18] It is not clear exactly when Scott started work again in Cambridge, but Dewar agreed to pay him £300 a year until the University came up with some permanent appointment, which was made eventually on 8 February 1892.[19] Moreover Dewar then added an additional £100 from his own pocket on condition that Scott 'will discharge such chemical duties as he may require'.[20] A Grace of the Senate on 4 September 1892 laid down that the Demonstrator's duties were to give 'catechetical [i.e. tutorial] or class instruction … and in supervising the practical work … under the general direction of the Professor'.[21] The additional £100 seems to make it clear that Dewar wanted Scott to continue the kind of outside commercial work to which Ruhemann had objected, although Dewar probably treated Scott in a more considerate way than he had treated Ruhemann.

Scott's second period in Cambridge was changed by the establishment of the Davy Faraday Research Laboratory as an adjunct to the laboratories of the Royal Institution in 1896, with Dewar and Lord Rayleigh as joint Directors. No doubt it was at Dewar's suggestion that they invited Scott to become the Superintendent of the new laboratory. Dewar's choice as his next Demonstrator in Cambridge was William Thomas Newton Spivey,[22] a Cambridge graduate with first-class honours in both parts of the Natural Sciences Tripos. From 1891 he acted as a private tutor and as a junior demonstrator in the Chemistry Laboratory. In 1896 he left to become science master at Epsom College, a school noted for the strength of its science side, but within the year Dewar recalled him and appointed him Jacksonian Demonstrator. He assigned Spivey to the same class that Scott had taught, that in advanced organic chemistry, 'in which subject his very wide reading and his extraordinarily methodical mind enabled him to arrange the vast amount of information in his lectures in an orderly, logical, and systematic manner which was of the greatest benefit to his pupils'. This assessment by his colleague and obituarist, T. Barlow Wood, could be read as a criticism of, and a contrast with, the less well organised lectures of his Professor. His research, carried out with Wood and Thomas H. Easterfield in the University Chemical Laboratory, was on the narcotic chemicals of Indian hemp, that is, compounds of the cannabis family. On 9 October 1901 he was using carbon disulphide as a solvent when the flask broke, the spilt solvent saturated his clothes, and caught fire. He died two weeks later, on 23 October, from the pneumonia that followed the severe burns.[22]

Spivey's successor was a very different man. Humphrey Owen Jones[23] graduated from the small University College of Wales at Aberystwyth in 1897 but, as was then often the custom for students from the minor British universities who wanted an academic career in science, he went up to Cambridge to take a second Bachelor's degree. Jones graduated with first-class honours in the Natural Sciences Tripos from Clare College in 1900. During his last year he had worked with a Demonstrator in Chemistry, H.J.H. Fenton,[23] on the oxidising reagent that now bears Fenton's name – a solution of hydrogen peroxide and a ferrous salt. Soon, however, Jones branched out into the field of stereochemistry, tackling the problem of the isomerism of the 'pentavalent' compounds of nitrogen, that is, of salts in which the four hydrogen atoms of the ammonium ion have been replaced by four different substituents. He was on hand at the time of Spivey's death and, two weeks later, on 8 November, Dewar appointed him Jacksonian Demonstrator. It was a good choice; Jones worked amicably with Dewar on the metal carbonyls, a speciality of Dewar's friend Ludwig Mond, he supervised his own students in his stereochemical research, and he undertook some investigations of the composition of Borneo petroleum. This last subject may have had some link to Dewar's own commercial interest in petroleum, and, indeed may have had some elements of the difficulty that had so blighted Ruhemann's career since on 26 March 1906 he had to apologise to Dewar for some offence at the same time as he accepted a cheque. Nevertheless Dewar came to treat him with more respect than he had shown to Ruhemann or, probably, to Spivey. Around 1905, when Jones was only 27, he raised the question of nominating Jones as a candidate for the Royal Society, but apparently thought better of it and did not complete the certificate until 26 October 1907. Jones was elected on 2 May 1912.

Dewar's health caused him problems in the early years of the new century and made him wonder whether he should give up the Jacksonian chair. Jones was formally nominated as his Deputy for the Lent Terms of 1906 to 1909. Liveing, now retired, wrote to Dewar on 21 November 1909:

> I am not surprised that you find the constant journeying backwards & forwards, to lecture here, too much. … If an outsider is to be brought in he needs to be a person of exceptional character, as well as of sufficient scientific merit … Of the men on the staff here now there is but one who could succeed you, I suppose, without much disturbance of the machine, namely Jones, who has been your deputy, and would go on in the same path as heretofore.

A few days later he concurred with Dewar in not trusting the present Vice-Chancellor and suggesting that Dewar defer his notice of retirement until the next October, when the Vice-Chancellor's term of office would expire.[24] In the end, Dewar's health improved and he changed his mind, and relieved Jones of his position as deputy. To the dismay of many in Cambridge, Dewar soldiered on until his death in office 14 years later.

Jones was an ardent mountaineer, a member of a group that gathered regularly at the hotel at Pen y Pass, at the head of the Pass of Llanberis, under the auspices of Geoffrey Winthrop Young, with whom he climbed in 1911 in the Mont Blanc range. The next year Jones married his student Muriel Edwards who shared his enthusiasm. They married on 1 August 1912 and were climbing above Courmayeur two weeks later when their guide fell and dragged all three to their deaths on the Fresney Glacier below. A memorial fund in Jones's memory was opened to subscription in Cambridge; it raised £3500 and led in 1914 to the endowment of the Humphrey Owen Jones Lectureship in Physical Chemistry. Jones's death was reported to the Council of the Royal Society on 3 October, who then asked Dewar, or, failing him, Charles Heycock, to write an obituary, but neither did so.[25] Dewar's involvement with Cambridge was now reduced to the merest formalities but he sought the appointment of one further Demonstrator, William Hobson Mills who held the post from Jones's death in 1912 until Dewar's in 1923. Mills was given the title as Dewar's deputy but the two men never worked together. Mills lectured on low temperature chemistry, a subject presumably chosen by Dewar. Mills's obituarist was to write: 'This subject could not have been of any great interest to him, nor could the conditions have been congenial, for Dewar on his very rare and brief visits to the University appeared to believe that on principle he should utter adverse criticism of his deputy's performance, of which he must have been in fact almost wholly ignorant.'[26]

Chapter 5
Spectroscopy

Liveing had long had an interest in spectroscopy. His colleague and obituarist, Charles Heycock, tells us that as a student Liveing attended lectures on physical optics and was the only member of the class to take up the lecturer's offer to demonstrate, after the lecture, the black absorption lines in a spectrum.[1] Before Dewar's arrival in Cambridge Liveing gave occasional courses on 'Spectroscopic Analysis'.[2] He had, however, at that time shown little interest in research and none in spectroscopic research. It was also a new subject for Dewar, who may have been looking for a more systematic field of research to replace the almost random set of topics that he had tackled in Edinburgh. The two men soon were soon on good terms in spite of what Heycock said of their 'widely different temperaments and widely different ideals' and of both being 'quick-tempered'. In 1884 Dewar, Liveing and the younger William Ramsay were on their way to the BAAS meeting in Montreal when Ramsay assessed Liveing as 'a nice old boy but not much of a chemist, as I should judge from conversation'.[3] Dewar's assessment was that Liveing's contribution to their joint work was 'thorough and trustworthy rather than original'.[4]

Chemists had had a keen interest in spectroscopy since Robert Bunsen and Gustav Kirchhoff at Heidelberg had shown in 1859–60 that the dark lines that Joseph von Fraunhofer had found in the spectrum of the Sun could be related to the same lines in the emission spectra of many elements in flames. Bunsen made use of such characteristic lines in 1861 in his isolation of the hitherto unknown metals rubidium ('red') and caesium ('blue'). He was quickly followed by Crookes and Claude August Lamy who independently found thallium, thus adding the 'green' element to the chemists' list of metals.[5] In 1863 two German chemists added indium, the metal with an indigo line.[6] There followed an outburst of chemical research and reviews of the new field. Henry Roscoe gave a Friday Evening Discourse at the Royal Institution on 1 March 1861 on the achievements of Bunsen and Kirchhoff, and the next year he gave there a course of three lectures on spectroscopic analysis.[7] Allen Miller, a photochemist at King's College, London, gave a well-received historical review of the field at the British Association meeting in Manchester in 1861, and the next year in a lecture to the Pharmaceutical Society.[8] Miller and the astronomer William Huggins extended the subject to an examination of the spectra of the stars from which they deduced that these contained many of the same elements found commonly on the Earth and in the spectrum of the Sun. Miller used this work as a basis for a short course of lectures on 'Spectrum Analysis, with its application to Astronomy' at the Royal Institution in May 1867.[9] And, as we have seen, in the 1870s, Liveing tried to interest the Cambridge undergraduates in the field. Throughout the 1860s and 1870s, optical spectra were the province of the chemists and the astronomers; most physicists took relatively little interest in a field that was teeming with interesting practical results but one that lacked any theoretical underpinning. When the Royal Society instituted the Davy Medal for chemistry in 1877 the first recipients were Kirchhoff and Bunsen. The experimental techniques of spectroscopy improved after 1870 with the introduction of dry plates which increased the resolution, of quartz prisms which extended the range of frequencies into the ultra-violet, and of H.A. Rowland's concave diffraction gratings which avoided the use of lenses.

Liveing and Dewar's joint work started with papers on the reversal of spectral lines; that is, the absorption of light from a continuous spectrum by the sharp lines of a metallic vapour leaving a 'black' line in the continuous spectrum. It is not clear how they divided this work between Cambridge and the laboratories of the Royal Institution, where Dewar worked after his appointment to the Fullerian chair in 1877. Most of it was probably done in London where the facilities were better, but their address is usually given as 'Cambridge' in joint papers, although Dewar used the RI for papers of which he was the sole author. John William Edward Heath, Dewar's assistant at the RI, said that Liveing came up to London for experiments and for the projection of the spectra in the RI lecture theatre from a fire-proof 'house' specially constructed for the purpose.[10] For the study of the inert gases Dewar prepared tubes in London which were sent to Cambridge for Liveing to study there.[11] The production of metallic vapours required high temperatures, which in turn meant heavy electric currents, either from 40 or so Grove cells or, from 1878, from a 'dynamo-magnetic-electric machine' that William Siemens presented to the Royal Institution early that year.[12] This machine was made available to Dewar 'by the kindness of Dr. Tyndall', a phrase that emphasises Dewar's then junior status in the Royal Institution. Norman Lockyer, who was to be their arch rival in their early years, obtained a Siemens machine the next year in order to attain the greatest temperature available in the laboratory. The Institution then acquired also a 'de Meritens' machine[13] which was less powerful than the Siemens machine but which Dewar also used. It produced an alternating intermittent current and so his attempts to use it for electrolytic experiments achieved little.[14] Some of the development of the spectrometers fell more to Liveing at Cambridge[15] and from 1893 he had there the assistance of the spectroscopist, John Edward Purvis.[16] There was a concave grating listed in the equipment of the chemical laboratory shortly after Liveing's retirement in 1908.[17] On the other hand, the RI had acquired a 'Krauss Universal Spectral Apparatus No. 7' in December 1898 at a time when Liveing and Dewar were embarking on their spectroscopic work at the temperature of the newly available liquid hydrogen.[18]

At first their work was purely fact-gathering, with no attempt at interpretation, but they were soon drawn into controversy with Lockyer. He had entered the field before them and at first their streams of short papers appeared independently in the *Proceedings of the Royal Society*, where a collision was inevitable. Lockyer came to believe that the so-called elements could be dissociated into something simpler at the highest temperatures of the Sun and of other even hotter stars. In 1872 Dewar was trying to calculate the temperature of the Sun, a topic on which he spoke at the British Association meeting in Brighton, and on 7 February 1873 he gave his first Friday Evening Discourse to the Royal Institution on the same subject.[19] In 1874 he lectured on dissociation before the Chemical Society.[20] The notion of the continuity of nature was a strong one in the nineteenth century;[21] if some molecules could be broken down into simpler entities by heating, or if a molecule like iodine could be separated into its atoms at even higher temperatures then, since there seemed no reason to suppose that temperature had any upper limit, it followed that even metallic elements should be capable of dissociation at some temperature. Where better to look for exceedingly high temperatures than in the Sun and the brighter stars?

Lockyer's ideas on dissociation were also imbedded in a wider area of chemical discussion that lasted throughout the nineteenth century.[22] These ideas were related to what can loosely be called the problem of 'Prout's hypothesis'. In 1815 and 1816 the physician and chemist William Prout[23] published two anonymous papers in which he put forward the related hypotheses that the atomic weights are integers, if expressed on the scale H = 1, and that all the elements are compounded from hydrogen atoms. The first proposal could be tested in the laboratory, although not without problems arising both from the difficulties of chemical analysis and the confusion at that time between equivalent and atomic weights. It attracted support at first from J.B.A. Dumas and his student, J.-S. Stas,[24] and opposition from J.J. Berzelius and J. Liebig. The lighter elements such as carbon, nitrogen and oxygen had integral weights, but some of the heavier ones, of which chlorine at about 35.5 was the most notorious, clearly did not. Stas made the accurate determination of the weights his life's work and by 1865 was convinced that the hypothesis of integral weights must be abandoned; it was 'une illusion pure'. It continued, however, to exert a strong influence on both physical and chemical thought, since the number that conformed to the hypothesis seemed to be too large for it to be due to chance. Could this feeling be quantified? There is no unambiguous way of doing this but in 1879 the Irish-born chemist J.W. Mallet, who spent most of his career at the University of Virginia, attempted the calculation[24, 25] by proposing that a weight reported to be within ± 0.1 of an integer be accepted as integral. He then calculated the probability of the observation that at least 10 of the 18 weights that he took to be the best determined fell within these ranges, if the weights were distributed at random; the result was 1/1098. Some years later R.J. Strutt[26] thought Mallet's procedure was somewhat arbitrary and determined instead the sum of the fractional amounts (irrespective of sign) by which each weight differed from the nearest integer. He then calculated the probability that this sum did not exceed the observed value of 4.37 for the (by then) 27 well-established weights. His procedure was an adaptation of Laplace's calculation of the chance that the orbits of the planets are as close as they are to a common plane around the Sun. Since his procedure gave extra weight to major departures from the integral rule, such as that of chlorine, his result was a higher probability than that found by Mallet, but it was still small, 1/680. Such probabilities counted as strong evidence for the mathematically competent such as Lord Rayleigh (Strutt's father), although, perhaps surprisingly, not for Stokes. Rayleigh said, in his Presidential Address to the Mathematical and Physics Section of the British Association in 1882, 'Some chemists have reprobated strongly the importation of *a priori* views ... Others ... consider that the experimental evidence against the simple numbers is of a very slender character, balanced, if not outweighed, by the *a priori* argument in favour of simplicity. ... The time has come when a re-determination of the densities of the principal gases may be desirable – an undertaking for which I have made some preparations'. As we now know, this 'undertaking' was to lead not to a confirmation of Prout's hypothesis but to the discovery of a whole new group of elements, the inert gases.[27] In 1887, even Stas, then towards the end of his life, remarked wistfully on the closeness of so many weights to integral values: 'Il faut croire qu'il y a quelque chose là-dessous.'[24] We now know that he was quite right – there was something under all this.

An interesting vignette is provided by a meeting of the Chemical Society on 15 January 1885, when T.E. Thorpe reported a new measurement of the atomic weight of titanium, a matter which was then in dispute.[28] His new figure of 48.01, on the scale O = 15.96 or H = 1, was clearly acceptable as an integral weight, but Dewar, a man for whom the facts were paramount, said in the discussion that the integer hypothesis must nevertheless be rejected, because there was the contrary evidence of magnesium, zinc and probably cadmium. He gave no figures for these but the modern values, on the scale $^{12}C = 12$ or O = 16, are Mg = 24.31, Zn = 65.37 and Cd = 112.40, with Ti = 47.90. He probably restricted himself to metallic elements since hydrogen was then often regarded as an honorary metal. J.H. Gladstone added that nevertheless many elements did conform to the integral rule but he made no mention of Mallet's calculation.[29] There the discussion, like so many of the time, apparently ended. Armstrong was present but said nothing, or at least nothing for publication.

It was the second part of Prout's speculation that underlay the later spectroscopic work. If all elements were compounded from hydrogen could their dissociation be detected in the spectra of the Sun and stars? A weaker form of the hypothesis was the observation that Mendeleev's Periodic Table of 1869 showed that there were relations between the elements that implied some common factor in their organisation, although not necessarily for the compounding of the elements from hydrogen. The exploration of the group of rare earths led to a similar conclusion, for here was a set of metals that were so similar in their properties that they surely must have some constitution or structure in common. Crookes[30] adduced both these forms of evidence in his Presidential Addresses to the Chemistry Section of the British Association in 1886 and to the Chemical Society in 1888. Lockyer tackled the problem spectroscopically, basing his argument on the observation that the solar spectra of different metals apparently had lines in common. This was not a new observation; the American astronomer C.A. Young[31] of Dartmouth College had compiled lists of solar lines from 1872 and had pointed out that if the common lines were not due to impurities then 'they were due either to a failure of the instruments of the day to resolve close lines or to some physical relationship, some molecular constitution perhaps, between the metals'. He called this possible relationship 'a resemblance, as regards the manner in which the molecules are built up out of the constituent atoms sufficient to establish between them an important physical (and probably chemical) relationship'. Lockyer first publicly declared his hand in his Bakerian Lecture[32] of 1873 when he discussed the evidence for the dissociation of both metals and metalloids (as the non-metals were then called) in the solar spectrum. He argued his case at a lively and well-attended meeting at the Royal Society on 12 December 1878, but his views received a cool reception from the chemists. The paper is a long one and the convoluted argument must have been difficult to follow orally, but Crookes at once reprinted it in his own journal, *Chemical News*.[33] He wrote to Lockyer a few days after the Royal Society meeting to say that he thought Lockyer had been badly treated in the discussion.[34] Roscoe was a sympathetic but sceptical friend who believed that atoms were probably divisible but he had attended the meeting specifically to emphasise that it was almost impossible to prevent the contamination of spectra with lines from impurities. Later he was to say that chemists should for the present continue to regard elements as the 'unalterable foundation stones' of their science.[35] Stokes wrote to Lockyer the next month to say that he supported his thesis that the elements were compounded – 'I have long supposed for my part that they were' – but he emphasised that Lockyer's evidence was not conclusive. It is perhaps surprising that he was not more impressed by the evidence of the frequency of integral atomic weights; 'considering the latitude to be left for errors of determination of atomic weights, and the large number of elements we have to play upon, the chances are that we could find trial cases of apparent coincidence'.[36] Would he change his mind a few months later when he had seen Mallet's calculation of the small probability that there was coincidence?

It was Lockyer's controversial claim[37] in 1878 that elemental carbon can be seen in the spectrum of the Sun's chromosphere that first led to public disagreement with the newly established firm of Liveing and Dewar. In Part V of their series of papers 'On the reversal of the lines of metallic vapours' they had given reasons why they did not accept this claim, and so opened an argument that ultimately went in their favour.[38] The argument became more personal when Liveing and Dewar accused Lockyer of relying on a '*confidential, uncorrected proof*' of their paper and not referring to the printed version. Lockyer could be as hot-tempered as Dewar and was apparently dissuaded from sending one letter by his friend and sometime colleague Chandler Roberts of the Royal Mint who provided Lockyer with specimens of pure metals.[39] Roscoe came down from Manchester to warn the Royal Society meeting in December 1878 of the impossibility of guarding totally against the presence of spectral lines from impurities, but Liveing and Dewar apparently played no part in the criticism. In 1879 Lockyer plunged in more deeply at the meeting of the British Association at Sheffield with a claim that he had found lines of hydrogen in the spectrum of sodium and other metals.[40]

He stopped short of publicly discussing the breakdown of the metal into hydrogen but that was the interpretation generally put on his publications. Roscoe, ever cautious, was unable to repeat the experiment,[41] and Lockyer's claim soon fell to the ground.

The next point of contention was Lockyer's claim that different metals had lines in common which, he averred, could not be dismissed as arising from impurities. Young had established that some of these lines could be resolved in spectrometers of good resolution[42] and the demolition was completed by Liveing and Dewar in a paper[43] submitted to the Royal Society on 12 May 1881. Dewar ended his Discourse on this topic with the firm conclusion:

> The supposition that the different elements may be resolved into simple constituents, or into a single one, has long been a favourite speculation with chemists; but however probable the hypothesis may appear *à priori*, it must be acknowledged that the facts derived from the most powerful method of analytical investigation yet devised give it scant support.[43]

Stokes was the Physical Secretary of the Royal Society from 1854 until his election to the Presidency in 1885 and throughout this period he controlled with great care the content of the mathematical, physical and even chemical papers submitted to the *Philosophical Transactions* and to the *Proceedings*. He was, for example, to decline papers from Crookes on spiritualism even when they purported to be framed so as to admit of tests of the factual soundness of the evidence for extra-physical communication. Perhaps more importantly, he was a continual source of advice and criticism of papers from Fellows whose mathematical and theoretical attainments he saw were rudimentary. In the trio of Lockyer, Crookes and Dewar he was faced with an argumentative group of ingenious experimentalists. Dewar had shown considerable mathematical ability at school and at Edinburgh University, but seems to have let his skill rust when he started on his research. An interesting example of Dewar's aversion to theory came at a discussion at a BAAS meeting on the kinetic theory of gases. Dewar carefully presented the views of his absent colleague Liveing but said nothing on his own account.[44] As well as this mentoring from Stokes he came, later in his career, to rely heavily on his brother-in-law, Hamilton Dickson, a Cambridge mathematician. Stokes read the papers of these three men as submitted and was unsparing in giving advice and pointing out possible errors of interpretation; all knew how much they owed to him and Stokes was one of the few men for whom Dewar had unstinted praise – his fellow Scots, Tait and Thomson, were never ranked so high in his estimation.[45]

One feature of the scientific life of Britain in the late nineteenth century was the important role of the British Association. Its annual meetings attracted attendances in the thousands, and its many Committees, appointed for one or more years, to report particularly on problems of standardisation and calibration, were essential for the technical advances in different branches of science and for international cooperation. Liveing and Dewar served on two spectroscopic committees, where they were joined by Rayleigh, their colleague at Cambridge.[46] It is not clear how much any of them contributed to the work of these committees; certainly they never wrote any of the reports submitted to the annual meetings of the Association. These were usually entirely factual, and often merely lists of lines observed by different spectroscopists, but such lists served as useful sources about who was doing what in the field at a time when such information was hard to come by. Liveing and Dewar published 78 papers between 1874 and 1892 and so must be regarded as major contributors to the field but the BAAS reports show that they were responsible for only a small part of the information put before the public. Between 1870 and 1884 the reports list about 2200 spectroscopic papers.[47]

In 1881 the British Association held its Jubilee Meeting in York, the city in which the first meeting had been held in 1831. The opening session was on Wednesday, 31 August, and the next day William Abney wrote to Lockyer to tell him to come quickly to York since he had heard the Dewar was to give a paper the next week in which he would seek to demolish Lockyer's views of the composition of the elements, and that he would be supported by his friends Armstrong and Schuster.[48] Dewar gave his paper on the following Monday; Lockyer was not present. Papers at BAAS meetings could be printed in full or in abstract, or could be recorded by title only. Dewar's paper appeared by title only: 'On the alleged decomposition of the elements'. Such treatment did not necessarily imply any criticism by the organisers; it may have been only that Dewar did not bother to submit a manuscript. More unusual was the failure to print Dewar's Address when he had been President of the Chemistry Section two years earlier in Sheffield. In both cases we know a little more about what he said from the reports from an unidentified journal and from *The Times*.[49] His York paper was, indeed, a criticism of Lockyer's position, and he was backed by Stokes who found his case convincing.[50] His Presidential Address at Sheffield had taken note of the industrial importance of the host city; 'steel, soda, and potash – these and similar topics were the burden of Professor Dewar's address'. In 1882 in Southampton it was Liveing's turn in the Presidential chair and the reporter from *The Times* expected a different approach:

> The two professors may, indeed, be regarded as the champions of the school opposed to Mr. Lockyer, and Mr. Dewar especially wages a somewhat bitter war against his fellow-worker in science. Professor Liveing is not, however, given to bitter speaking and while he may be expected to survey the various lines of research that have a bearing on the hypotheses referred to, he is likely to do so with the dignity and impartiality of a seeker after truth.

This he did.[49] Rayleigh viewed spectroscopy with mixed feelings. He was later to appreciate its power when it came to the separation and identification of the inert gases at the end of the century, but, notwithstanding his co-option on to the BAAS committees, he seemed less enthusiastic in his earlier days. He thought that the field was one of an infinity of facts and an almost total lack of their interpretation.

He was President at the 1884 meeting of the British Association in Montreal, a meeting which Liveing and Dewar both attended, and so presumably were in the audience for his Presidential Address.[51] Here he summarised the recent advances in the physical sciences, saying of spectroscopy: 'The investigation of the relation of the visible and ultra-violet spectrum to various forms of matter has occupied the attention of a host of able workers, among whom none have been more successful than my colleagues at Cambridge, Professors Liveing and Dewar. The subject is too large both for the occasion and for the individual, and I must pass it by.' Much later in the Address he may have had such work in mind when he said that 'Detached and ill-assorted facts are only the raw material, and in the absence of a theoretical solvent, have but little nutritive value'.[51] A later comment came in October 1891 when he wrote that 'Dewar has tempted me to try the spectrum of my hydrogen. I hope it won't be the beginning of an immersion in a bog of spectroscopy'.[51]

In Montreal Ramsay, certainly, and Dewar, probably, met Willard Gibbs. They had both been recipients of reprints of his first two papers in 1873 and 1875 on thermodynamic surfaces which Gibbs distributed widely, although neither was sent the massive paper on 'The equilibrium of heterogeneous substances' of 1878–9 which Gibbs sent only to those whom he perceived as having strong mathematical skills. When Maxwell received the first two papers he prepared three-dimensional plaster models of an Entropy-Volume-Energy surface for a substance like water in which the density of the liquid was greater than that of the solid. These he sent to many physicists and physical institutions in 1875–6, including the RI. That an early one went to Gibbs is implied by a letter to him from Alexander Freeman of St John's College, Cambridge on 18 February 1875, but he was still making and offering models as late as December 1877. It is, however, hard to know what to make of an assertion of the mathematician Joseph Sylvester in a letter of 12 June 1880 that Dewar had told him that he had seen Maxwell preparing to send a copy to Gibbs in October 1879, only a fortnight before he died. Maxwell was then an ill and very weak man who had returned from his Scottish estate to Cambridge on 8 October knowing that he had only weeks to live. He died on 5 November and is unlikely to have been preparing such a gift at that time. In the same letter Sylvester emphasised Dewar's great respect for Gibbs's work, for when Sylvester said that he supposed that Gibbs was the best man in America in this line of work, Dewar had replied 'Aye – and in this country too'.[52]

After the Montreal meeting the British delegates were invited to the meeting of the American Association in Philadelphia; over 300 made the move, encouraged by a subsidised railroad fare of about £1 for the journey. Dewar was an official representative here from the Royal Institution and he contributed papers[53] on graphical formulae and their misuse, and a measurement of the density of solid carbon dioxide. Some of the caravan then moved to Baltimore where Thomson was to deliver an inspiring, if at times eccentric, set of theoretical physics discourses, known ever since as the Baltimore Lectures.[54] Rayleigh and the electrical engineer George Forbes, the younger son of James David Forbes, were among Thomson's

21 auditors but Liveing and Dewar, no theoreticians they, were not tempted. Dewar, armed with 'no end of letters of introduction', spent the remainder of his visit exploring the American petroleum industry.[55] His involvement with the Nobel brothers (see Chapter 7) centred at that time on their common interest in the Baku oilfields in the Caspian Sea.

The first major work of Liveing and Dewar to appear after their series of bouts with Lockyer was an extension of their cataloguing of the lines of the elements into the ultra-violet, aided by the availability of an accurately-ruled grating with 17296 lines to the inch, made by the American amateur astronomer Lewis M. Rutherfurd. This work appeared in 1883 but was a summary of experiments that had extended over the preceding three years.[56] Thereafter their papers became less frequent and were often on the use of spectroscopy to throw light on other lines of research such as gaseous explosions, on Dewar's increasing interest in the liquefaction of gases, and on other miscellaneous chemical and astronomical topics. There was a two-year break between 1884 and 1886 when they published nothing. Their work now often appeared in the *Philosophical Magazine* and other less austere journals than the *Proceedings* and *Transactions* of the Royal Society. Both men sometimes published separately and it became clear that Liveing was the more inclined to take risks with occasional attempts at theoretical speculation, as, for example, when he doubted whether there was any valid reason why the energy within a molecule should be distributed evenly over all degrees of freedom, or whether a ratio of the specific heats of mercury vapour, C_p/C_v, of 5/3 was proof that the vapour was monatomic. The liquefaction of hydrogen in 1898 gave them a new opportunity for studying spectra down to even lower temperatures.

Throughout all their joint work they kept close to the facts that they observed. Naturally the apparent regularities in the appearance of elemental spectra attracted many physicists into attempts to fit these wavelengths by numerical expressions. Schuster described some of these attempts to the British Association[57] in 1882, but the breakthrough came from a hitherto unknown Swiss schoolmaster, Johann Balmer,[58] in 1885. He showed that the lines in the visible part of the spectrum of atomic hydrogen could be represented by wavelengths proportional to $m^2/(m^2-2^2)$, where $m = 3, 4, 5$, etc. It was soon found that it was simpler to use the reciprocal of the wavelengths, that is, the wave numbers, or number of lines per centimetre, as the measure the characteristic distribution of lines in a spectrum. Balmer's result then requires that the distribution becomes proportional to $(1/n^2) - (1/m^2)$, where for the series of lines in the visible part of the hydrogen spectrum, $n = 2$, and where, as the Swedish spectroscopist Johan Rydberg[59] and others then showed, there are also series of lines elsewhere in the hydrogen spectrum that could be fitted with $n = 1$, $m = 2, 3, 4$ etc., with $n = 3$, $m = 4, 5, 6$ etc.; and with $n = 4$, $m = 5, 6, 7$ etc. Such simple expressions did not fit other atomic spectra but the alkali metals, lithium, sodium and potassium, had certain hydrogen-like traits. Liveing and Dewar played no part in this numerology and it is therefore ironic that, when the so-called 'old quantum theory' was developed in the early twentieth century, the purely empirical description that they had used in their papers became the basis of the quantal classification of atomic levels. Quantum theory requires that the wave number, or, what is equivalent, the frequency of a line, measures the difference of energy between two states or levels that the atom can take up; these levels are described by the numbers m and n above. Rydberg adopted certain adjectives that Liveing and Dewar had used to name these lines as 'sharp', 'principal' and 'diffuse'. The first and third of these adjectives are purely descriptive of the appearance of the lines and were used repeatedly by Liveing and Dewar in their papers on the alkali metals and elsewhere. The second adjective, 'principal', could carry some theoretical meaning but its relatively rare use by them seems to imply nothing more than that the series of lines was prominent.[60] A fourth series in the spectra of the alkali metals was later given the name 'fundamental' by Arno Bergmann[61] in his Jena thesis of 1907. This description is clearly theory-laden and so not directly comparable with Liveing and Dewar's descriptions of 20 to 30 years earlier. All four names, usually in the guise of their initial letters, *s*, *p*, *d* and *f*, have now passed into quantum theory to denote the values of the units of angular momentum of electronic motions in the atom, namely $l = 0$, 1, 2 and 3. Fanciful naming has been carried no further and if higher numbers are needed then $l =$ 4, 5 etc. are now simply represented by the next letters of the alphabet, *g*, *h* etc. It is not clear why Bergmann thought that the level $l = 3$ is 'fundamental' but one possible explanation is that states with high angular momentum have wave-functions that are small near the atomic nucleus,[62] which is the quantal expression of the classical condition that high angular momentum flings out the orbiting electron well away from the nucleus. The low electron density near the nucleus leads to the spectral lines forming a series that is close to that of a hydrogenic atom and so simpler in spacing than those from the *p* and *d* states. Be that as it may, it is ironic that the strictly pragmatic work of Liveing and Dewar has left us a legacy in quantum theory, a subject with which they seem to have had no concern whatsoever.

Chapter 6
London

The Royal Institution of Great Britain was founded in 1799 for applying science to 'the common purposes of life', a brief that came in the nineteenth century to encompass a flourishing research school in the physical sciences and a strong programme of lectures for popular instruction and for the dissemination of science.[1] Its government was unusual, with an executive board of Managers elected from and by the members, and a board of Visitors, elected similarly but whose non-executive role was simply to report to the members on what the Managers were doing and to comment on their decisions. The officers were the President, Treasurer and Secretary, of whom the first had only formal duties unless there was a crisis in the affairs of the Institution. This office had become almost a heredity right of the Percy family in the persons of the successive Dukes of Northumberland. From 1842, the year of Dewar's birth, until 1930, seven years after his death, the post was occupied by the 4th, 6th, 7th and 8th Dukes, with only a break of eight years from 1865 to 1873 when it was held by Sir Henry Holland, Queen Victoria's physician. The Managers met on the first Monday of each month from November to July, and reported to a Monthly Meeting of Members later the same day. The Institution has always been situated at 21 Albemarle Street in London's West End. Its early years were associated with the work of Humphry Davy whose spectacular research on the chemical elements established its scientific reputation. A constant supporter was John Fuller, the owner of a plantation in Jamaica and sometime Member of Parliament. In 1833 he endowed a chair of chemistry for the benefit of Michael Faraday, who became then entitled the Fullerian Professor of Chemistry. A few months later Fuller endowed also a Professorship of Physiology. Both chairs were established as three-year posts, with the opportunity of renewal, but it was only the chemistry chair that became accepted as a post that a man could expect to occupy for his active life. The chemist usually attended the meetings of the Managers while the physiologist did not. Fuller's gift was £10000 in 3 per cent Consols and the dividend was split into three parts, one third, or £100 p.a., went to the chemist, that is to Faraday, one third to the physiologist, and one third to be left to accumulate until it was worth £10000, a target that was reached in 1865. There was then £16666/13/4 to divide, so each professor received £250 from the trust fund. Faraday's senior colleague was the chemist William Brande, and when he retired in 1852 he was succeeded by John Tyndall[2] with the title of Professor of Natural Philosophy. The cost of the salaries of Brande, Tyndall and Edward Frankland,[3] who held a Professorship of Chemistry for five years from 1863 to 1868, was carried by the Royal Institution itself. Faraday and Tyndall together made the RI (as it became popularly known) one of the leading scientific and social institutions in London. When Faraday retired in 1867 Tyndall became the Resident Professor, with the title of Superintendent of the House, and William Odling was offered the Fullerian chair.[4] He remained in post until he was elected Wayneflete Professor of Chemistry at Oxford in 1872, part-way through his second three-year appointment. His departure in 1873 left the Fullerian chair vacant and there seems to have been some doubt as to what the RI should then do, presumably because there was no obvious successor. Heinrich Debus, a friend of Tyndall from their days at Marburg, and Henry Armstrong were said to be candidates.[5] John Hall Gladstone[6] was then a Manager but he retired from this post to offer himself as a candidate, and was elected to the chair on 4 May 1874. He was two years older than the departing Odling. He was given the title of Director of the Chemical Laboratory, a name that was to cause some difficulty later since Tyndall was Superintendent of the House and Resident Professor. Gladstone's appointment was apparently not intended to be a permanent replacement of Faraday; the Managers recorded in a minute of 6 March 1876 that 'Dr Gladstone does not wish to withdraw from the understanding that he would relinquish the professorship, when an approved successor had been found'.[7] He asked to stay on for some time after that date since he had some research that he was anxious to complete. He and Tyndall did not get on well together, for Gladstone was an evangelical Christian who was upset by the agnostic nature of Tyndall's notorious Presidential Address to the British Association in Belfast, a few months after Gladstone's election.

In February 1877 the Managers authorised the publication of an advertisement:

> The Fullerian Professorship of Chemistry will be vacant on 4th of May next. Candidates are requested to apply in writing before March 31 next.
>
> William Spottiswoode, Secretary.[8]

The Committee for making the appointment was chosen to be: 'Sir Frederick Pollock, Dr Hooker, Mr Siemens, the Treasurer [George Busk] and the Secretary.'[9] Hooker, Busk and Spottiswoode, together with John Tyndall, were members of the nine-man X-Club, a dining club that exerted a strong behind-the-scenes influence on British science from 1864 until its break-up through the old age of its members in 1893.[10] The Club, of which Huxley was the leader, was liberal in political and social matters, firmly in favour of fundamental research, and dedicated to the reform of the Royal Society.[11] There is no reason to think therefore that its members would have any bias in favour of Dewar whose politics and general attitude, at least in later life, were conservative.

Dewar's formal duties at Cambridge were less than a full-time job and he was unhappy with the lack of facilities for research. The London post was one that he could hold in parallel with the Cambridge chair, the laboratory facilities were potentially more extensive, and the extra salary was attractive. He had already given three of the renowned Friday Evening Discourses, which he probably found to be more suited to his style than the undergraduate lectures he had to give at

Cambridge, and he had agreed in April to give three lectures on Davy,[12] for he had a strong interest in the history of chemistry. He therefore applied for the vacant post. When the applications closed he had been joined by two men with whom he had been in competition at Cambridge – Armstrong and Hartley. The other candidates were: John Buchanan, Dewar's former colleague as an assistant in Edinburgh; William Ramsay, another Scots chemist, then an assistant at Glasgow University, with whom Dewar was to wage a famous war 20 years later; George Rodwell, a chemist and traveller who was the Science Master at Marlborough College; Alfred Tribe, Gladstone's assistant and Science Lecturer at Dulwich College; and C. Alder Wright, a chemist at St Mary's Hospital who had recently synthesised heroin.[13] The actual group that made the appointment was Busk, as chairman, Pollock, Siemens, Spottiswoode and Frankland,[14] another member of the X-Club, who had been co-opted. Hooker missed the meeting; perhaps he was too busy with the preparations for his departure in June for a major expedition to the Rocky Mountains with Asa Gray.[14] The Committee reported on 2 April 1877; a week later the Managers, after a ballot, elected Dewar to the Fullerian Profesorship.[15] His salary was set at £250 from the Fullerian dividend plus £300 from the funds of the RI. This total was the same as Tyndall was paid but, as Resident Professor, he received also the benefit of free board and lodging. Both received also the fee of £10 paid for each lecture, the same sum as was paid to outside lecturers. The long letter of Dewar's appointment carefully avoided naming him as Director of the Chemistry Laboratory; the Managers did not want a repetition of the argument about Gladstone's retirement. The previous holders of RI chairs had all been appointed as promising young men – Davy, Brande, Faraday and Tyndall. Dewar, in common with his fellow applicants, continued in this tradition; he was 34 at the time of his election. Thereafter the pattern changed and the incoming professors were more senior men who had already made their name outside the RI at the time of their election – Lord Rayleigh, J.J. Thomson, Ernest Rutherford, William and Lawrence Bragg, Edward (Percy) Andrade and George Porter. By chance, at the same meeting as Dewar was elected to the chair, Rayleigh became a Manager.

In June Dewar attended the Managers' Meeting for the first time and a committee was set up to enquire into the state of the chemical laboratory. It comprised Warren de la Rue (a Manager), Tyndall and Frankland; Dewar was not a member but was invited to attend.[16] Whatever shortcomings the committee revealed were remedied, at least in part, by the gift early the next year[17] of £50 from de la Rue, and of a steam engine and a dynamo-magnetic-electric machine from Charles William Siemens, gifts which will be discussed later. Meanwhile Dewar got his teaching off to an ambitious start with a course of 12 lectures on 'The chemistry of the organic world', which he agreed to give between January and April 1878, and a Friday Evening Discourse on 'Experiments in electro-photometry' on 29 March.[18] There is no record of how these were received but there is no doubt that he soon adapted his style to what had become the RI custom of impressive facts illustrated by ingenious and often spectacular experiments. This tradition goes back to the earliest days of the Institution and, indeed, continues to this day in the Friday

Evening Discourses. Bernard Becker has given a lively account of a successful afternoon lecture at the RI, after he had heard a competent but dull and thinly-attended lecture by a visiting speaker. The one that Becker chose to describe was by Tyndall and took place three years before Dewar's arrival, but his performances eventually rivalled those of Tyndall. Becker wrote:

> On another afternoon, about 3 p.m., I betake myself to Albemarle Street, and become the spectator of a widely different scene. The theatre is already full of eager visitors and thirsters after science, when elucidated by those brilliant experiments which excite the admiration and envy of Professor's Tyndall's imitators. ... At a three o'clock lecture many ladies are, of course, present in all the variety of gorgeous array now in fashion ... Do they know much about liquids and gases, or have they come to learn? Verily, I know not. The well-known lecture table is covered with apparatus, and a huge bath tub occupies a considerable space. Mr. Cottrell, the laboratory assistant, is very busy. ... The lecture, interesting in itself, is rendered doubly so by numerous and beautiful experiments. ... The prescribed hour appears unnaturally short when the clock strikes, the lecture is closed by a short sentence and, amid a mighty rustling of silks, the audience prepares to depart. For a few minutes a talkative crowd blocks up the wide staircase and hall, and a sort of scramble takes place for the carriages of which Albemarle Street is full. Fashion takes its departure and, having had in science enough to last a week, leaves the professor to enjoy himself in his admirably-appointed laboratory.[19]

Dewar took his RI duties seriously; during the 45 years in which he held the Fullerian chair he gave 49 Friday Evening Discourses, one Commemorative Discourse, eight sets of Childrens' Lectures at Christmas, and 39 courses of afternoon lectures. He continued the tradition of Faraday and Tyndall of inviting distinguished men from a wide range of pursuits.[20] His own lectures started well but did not always rise to the standard that Becker claimed for Tyndall, who wrote to Spottiswoode after Dewar's first year in post: 'I am greatly rejoiced to hear of the success of Dewar's lectures. He is a genuine man; and has had to endure a good deal at the hands of men who were the reverse of genuine.' It is not clear if Tyndall is here criticising Dewar's Cambridge colleagues or his Edinburgh mentor, Guthrie Tait. Tyndall's enthusiasm, however, did not last unalloyed for, even after seven years of Dewar's lecturing, he could only grudgingly praise his efforts: 'Heard Dewar lecture. He is improving very much, though he still does not resist the temptation to bring in experiments which do not belong to his exposition.'[21] He had at least one failure; on 5 June 1885 he gave a Friday Evening Discourse on his preparation of liquid air; it was not the first time in Britain that liquid air had been prepared and shown to an audience – that he had done a year ago[22] – but it was still a great novelty. The mathematician Thomas Archer Hirst described the occasion as 'A more *noisy* and *unsatisfactory* lecture I never was present at. The noise was caused by the compression pumps all around me, and the unsatisfactoriness by the failure of the principal experiment and the absence of any thing worthy of the name of scientific exposition'.[23] It is to be hoped that Dewar had attended to at least some of these faults when he repeated the lecture three weeks later for the Prince and Princess of Wales. Eventually, however, he established a reputation as a fine public lecturer whom the Managers of the RI valued highly for his ability to bring in large audiences. It was, perhaps, unreasonable of Hirst to expect an account of the thermodynamics of liquefying air in a popular lecture of the kind that a Friday Evening audience expected, but Dewar's later more technical lectures and papers were also marked by the absence of any serious discussion of underlying physical theory. But when things went well, his spectacular style, and the wealth of the dramatic and generally successful experiments he used as illustrative material in his Friday Evening Discourses, ensured that everyone left the theatre feeling that they had spent the hour wisely. McKendrick's verdict was, 'As a lecturer he was splendid, as you know, but his fault, if any, was that he prepared far too much for the hour, and had too many beautiful experiments'.[24] But we often do not see ourselves as others see us, and after Dewar had had to sit through an excessively long Bakerian Lecture by William Crookes at the Royal Society he said the next day to Walter White, the Assistant Secretary, that 'a scientific man should never exceed twenty minutes in an exposition; that he knows by experience that much can be spoken in that time'.[25]

John Cottrell was Tyndall's lecture-room and laboratory assistant. Dewar soon felt the need for an assistant of his own to match the help he had in Cambridge from Alexander Scott. The Managers agreed, and in 1878 he chose a 25-year-old chemist, Gerrard Ansdell.[26] Dewar allowed him more freedom of action than was to become his custom with his assistants in later years. He encouraged Ansdell to embark on a research programme of his own, using the 'Cailletet pump' that had been given to the RI by Warren de la Rue.[27] Dewar, who had recently been elected to the Royal Society, probably took pride in submitting Ansdell's three papers for publication in the *Proceedings*. They were on the physical properties of gases and liquids at elevated pressures, a branch of research in which Dewar had no experience but which he proposed to Ansdell as one related to Dewar's intended move into the related subject of the liquefaction of gases at low temperatures, a field in which, as he was well aware, the RI had had a great reputation in Faraday's day. Ansdell's work added little to the studies by Thomas Andrews and James Thomson in the early 1870s on the condensation of gas mixtures at around their critical points.[28] The Cailletet pump was modelled on the apparatus of Andrews, and so was well-adapted to Ansdell's work. Dewar's second Friday Evening Discourse after his appointment at the RI was a largely historical lecture on 'Liquefaction of gases', on 14 June 1878, a subject on which he also spoke at the meeting of the BAAS later that year.[29]

In December 1880 Ansdell obtained leave of absence and went with one of his brothers on a voyage to seek out their eldest brother, Richard, who was believed to be farming in Fiji. Gerrard spent five months there, writing a book on the principal island and its inhabitants which he illustrated with 44 of his own photographs; the glass plate of the 45th was broken before he could get it back to London. This book is now extremely rare; a copy found in an Oxfam bookshop was auctioned for £37000 in April 2010. Photography was an interest of Dewar's but it never became a serious part of his work. It was, however, a popular subject at the RI in the 1880s and 1890s. Ansdell was replaced as Dewar's assistant for a few weeks by Alexander Scott coming in from Cambridge. While Ansdell was away Dewar used the Cailletet pump for some qualitative experiments on the effect of added gases and liquids on the critical state of carbon dioxide. This work required pressures of over 100 atm, possibly Dewar's first direct experience of the problems of this field.[30] On Ansdell's return to London, he can have spent little more time in the employ of the RI since in December 1881 he was replaced by Robert Lennox who was to play a major part in Dewar's work on the liquefaction of gases. Ansdell then set up as an independent consulting and analytical chemist. He did, however, manage to complete further work with Dewar who joined him as co-author in a fourth paper in 1886.[31] This involved much analytical work on the gases found in meteorites and it is not clear whether this analysis was carried out at the RI or at premises that Ansdell must have established in order to further his independent career. The paper ends with the intention of further collaboration – 'we propose continuing this investigation' – but nothing more was done together.

There soon arose a problem with Dewar's RI appointment, apparently from its interfering with his duties at Cambridge. As his first three-year period came to an end, he wrote to the Secretary on 27 February 1880:

> Dear De La Rue,
>
> My Fullerian Appointment is now virtually ended and after much consideration, I desire to place my resignation of this Office in the hands of the Managers of the Royal Institution. This action arises solely on account of personal considerations touching work, health etc and for no other reason.
>
> Yours faithfully, James Dewar[32]

Whatever the trouble was, it was soon resolved after the Managers had asked de la Rue and Tyndall to discuss the problem with him. On 1 April he wrote again to say that 'the kind interviews I have had with Dr Tyndall and yourself have induced a withdrawal of my resignation'. He was re-elected for the next three-year spell a week later.[32] He signalled his commitment to the RI by becoming a Member, although it was another eight years before his wife joined, when they moved to London on his becoming Superintendent of the RI after Tyndall's retirement. He was to remain in the two posts at Cambridge and the RI for the rest of his life. He made an application for the chair at the Royal College of Science in 1885 but he was not appointed; Thomas Edward Thorpe was the man chosen.

His relations with Tyndall were never easy but were formally correct. Both men had difficult temperaments, and Tyndall had long been at odds with Dewar's former mentor and collaborator, Guthrie Tait, over the theory of the flow of glaciers and on their differing understanding of thermodynamics and its history, so he probably approached Dewar warily. Dewar lived in Cambridge for the 10 years that he and Tyndall overlapped in their posts at the RI but he often stayed or dined with the Tyndalls when he came up to London to lecture or to start his major research programme on the liquefaction of gases. After the trouble over the precise position that Gladstone had occupied as Fullerian Professor, the Managers now ensured that Tyndall's seniority and responsibility were always made clear. For example, the committee that had looked into the state of the chemistry laboratory after Dewar's appointment had not made him a member but had only invited him to attend; the gifts of Siemens and de la Rue for work the liquefaction of gases were to be used 'as Dr Tyndall may direct' although it was Dewar who was to carry out the work; when a second assistant to Ansdell was needed in the chemistry laboratory in 1879 Dewar appointed William Card 'with the consent of Dr Tyndall'; and when later in that year Spottiswoode resigned as Secretary on being elected President of the Royal Society on 30 November, Tyndall but not Dewar was on the committee to choose de la Rue as his successor. Spottiswoode remained a Manager, as Kelvin was to do later when he became President of the Royal Society.[33]

The ventilation of the lecture theatre at the RI, and the safety of its exits in the event of fire, had always been a source of worry and in 1884 Tyndall and Dewar had some disagreement on how these problems, and another with the drains, should be tackled. These matters were soon resolved but Tyndall thought Dewar's way of dealing with the problem of the ventilation of the theatre was inept; he wrote in his journal:

> Attend a meeting of the Building Committee in the afternoon. I was somewhat hurt at the manner in which Dewar spoke about the theatre. He knows little of the labour that I have expended on it. He had been testing, it appears, with the flame of a taper, but an indraught that would affect a flame would not be borne by the audience. Desired Cottrell to test it afterwards by the smoke of brown paper, and he found both indraught and outdraught perfectly normal. Dewar had asserted with some vehemence that not a particle of air entered the theatre. If he believed *such* to be the case he ought, I think, to have mentioned the matter to me in the first instance. I have done my best to make Dewar's position at the Royal Institution harmonise in every way with his desires. He is an honest and straightforward fellow, but permits himself from time to time to indulge in extreme assertions.[34]

The problem of lighting the theatre was solved in April 1882 when Sir Joseph Swan offered to install electric light at his own expense. It was found that the existing Siemens engine was not strong enough both for this duty and for the calls made on it by the research laboratories. Tyndall was authorised to buy a new one from Crossley Brothers for not more than £300, to which de la Rue contributed £100, Tyndall and Colonel James Grant (a Manager) £50 each, and Lachlan Rate (a Visitor) £20. Arguments with Crossleys over their bill continued for some years. A committee of Bramwell, Frankland, Bowman (the Secretary), Tyndall and Dewar was set up and asked to consider the matter 'and report thereon'. Richard Crompton, an electrical engineer and the founder of the electric light company named after him, was called in, and asked Siemens Brothers 'to exchange the large Dynamo Machine for one suitable for the power of the new engine'.[35] Cottrell again managed to upset both professors with his handling of the lighting. Tyndall wrote in his *Journal*:

> Cottrell mentioned to me that Dewar was 'furious' about the lighting of the theatre. Went & talked to him. Cottrell himself is, I must say, somewhat to blame. He ought not and need not have allowed the circuit in the theatre to be disconnected. Arranged for Mr Faraday to come & connect wires tomorrow morning. Dewar wrote a curious little note to me afterwards. He gets overstrained & flurried when hard worked. Cottrell again gave him a wrong impression, causing him to think that I had overheard him in the theatre & had therefore made inquiries about his anger. I did neither the one nor the other.[36]

Concern for safety required changes in the theatre that reduced its capacity. Popular lecturers such as Tyndall's friend, Dean Stanley, John Ruskin and some of the scientists, such as Tyndall himself, Thomas Huxley and later Dewar, had attracted audiences of over 1000, but new exits and the rearrangements of the seating reduced the capacity to 452 by the middle of the twentieth century.[37]

Tyndall's health was failing during much of the last decade of his tenure of the post of Superintendent of the House so Dewar had on occasion to assume more of the administrative duties than Tyndall would have wished. In June 1881 the Managers told Tyndall that he was 'suffering from past overwork' and that he should 'take a complete rest for ~~six~~ twelve months'. He thanked them for their consideration but said that he still had to complete some work for the Bakerian Lecture he was to give to the Royal Society the next year, but he was glad to be off to the Alps for the summer, his unfailing remedy for any problem with his health. He did, however, ask to be relieved of the Christmas Lectures for that year which were given by Robert Ball, the astronomer.[38] A more serious difficulty arose in 1886 when Tyndall's health again gave way. Dewar took on duties that should have been the responsibility of Tyndall[39] and uncovered some 'irregularities' in the conduct of affairs which he laid at the door of John Cottrell, with whom, as we have seen, he had never been on good terms. The only specific offence mentioned seems to be an earlier comment on Cottrell's excessive fondness for beer with his lunch.[40] The Managers, led by Sir Frederick Bramwell, the recently elected Secretary and a supporter of Dewar, responded by dismissing Cottrell, apparently with Tyndall's consent, although after he had retired he was to employ Cottrell privately to help him sort his papers and books. Tyndall was at first glad of Dewar's support in his illness and wrote to his X-Club friend Herbert Spencer to say that Dewar 'was a most noble prop to me throughout the ordeal. I would give a good deal of dross, which is most highly prized in the world, to have seen it continue so to the end. But it seemed to me that at the last he presumed somewhat upon the freedom which my absolute trust in his brotherly affection caused me to allow him, and this to the present hour is a source of sorrow to me'.[41] Nevertheless the Dewars dined with the Tyndalls on 10 August, the day before they left for their usual visit to the Alps. The Managers thought that Tyndall's health was again too precarious for him to be asked to give the Christmas Lectures at the RI, although it was his 'turn' and they asked Dewar to stand in. Tyndall and his wife were both upset by this decision and asked for it to be reconsidered but the Managers would not change their minds. Tyndall accepted their decision 'rather than introduce a jar where for 33 years I have lived without jars'. His health did not improve and the next year, on 7 March 1887, he submitted his resignation.[42] He offered a cheque for £200 to cover the duties he had not been able to carry out but this the Managers refused. He declined a pension. Dewar was at once chosen to succeed him as Superintendent of the House but again he upset the Tyndalls: 'the end has not been pleasant' because of the 'unseemly haste of the newcomers'.[43] Tyndall's retirement was marked by a grand dinner for 200 guests on 29 June at 'Willis's Rooms', arranged by Norman Lockyer and presided over by George Stokes, now the President of the Royal Society. Tyndall asked Hirst to see that Dewar, 'with whom he is not, at present, on friendly terms', did not have a prominent part in the proceedings.[44] Tyndall was elected an Honorary Professor at the RI but was disappointed not to be given any laboratory space there.

Two weeks after Tyndall's resignation Dewar was appointed Superintendent and was asked 'to occupy the House at such times as may be convenient to him'.[45] This liberal interpretation of his duties as Resident Professor and Superintendent of the House recognised his continuing responsibilities in Cambridge, and shows how highly the Managers thought of him as an asset that must be retained by the Institution. The Secretary, Treasurer and Dewar were at once asked to approach Lord Rayleigh, now no longer a Manager, to offer him the post of Professor of Natural Philosophy. He accepted on 2 April on the understanding that he could continue his research at his private laboratory in the family home at Terling in Essex.[46] He had already resigned his post as Cavendish Professor of Physics at Cambridge. It was agreed that his research should remain at Terling, and his salary was fixed at £300 a year with a grant of £100 to pay for an assistant who would work both at Terling and at the RI. After his appointment Rayleigh attended the Managers' Meetings less regularly than Tyndall and Dewar did, but it is characteristic of those formal and status-conscious times that, when he did, his name in the minutes preceded Dewar's, in spite of their alphabetical order and of Dewar's seniority in position and service. There was at first little overlap in the interests of the two men in spite of the fact that the title of Rayleigh's post at the RI was almost the same as that of Dewar's at Cambridge. Their relations were formal and correct although it was clear that privately Rayleigh must have found him an exasperating colleague. His son was to say that the Dewar was the only man who could make his father angry. Their interests, but not their relations, became closer after Rayleigh's discovery in 1894 of argon in the atmosphere (Chapter 9).

Tyndall retired to his house, Hind Head (or Hindhead) at Haslemere in Surrey, where he hoped to continue with some of his scientific writing and possibly with some experimental work in the RI. Shortly after he had retired he wrote, on 5 November 1888, to the Managers about the disposition of some equipment that had been acquired by him and which was now either left at the RI or was to be taken to Haslemere. One sentence was to cause trouble: 'The large Nicol prisms on the Royal Society table, I retain.' He found that there was an 'impression' circulating that he had earlier given these to the RI and so he was asked to come the Managers' Meeting on 5 December to sort out the question of ownership. He wrote to Hirst: 'I propose answering the summons to the Managers' Meeting tomorrow by appearing before them with a mind perfectly open to correction if I have erred in supposing the prisms to be mine.' When he arrived he found that the attendance of Managers was 'scanty' (nine were present), that de la Rue was in the chair and that Pollock, Bramwell and Dewar were there. From the discussion it was soon established that none of the Managers was responsible for what he thought was the false 'impression'. The clear implication was that Dewar was responsible although Tyndall does not name him directly even in the privacy of his journal. The minutes of the meeting report in guarded language that 'Professor Tyndall attended and made a verbal communication. Whereupon on the motion of M^{r} De La Rue, seconded by Sir Frederick Pollock, the Managers came to the conclusion that the Nicol prisms were not included in the gift of his part of the apparatus'. His position thus vindicated, Tyndall at once gave the prisms to Rayleigh, who was not present, but who was in a better position than Dewar to make use of them. He wrote in his journal that 'I then quitted the room probably leaving Sir F. Bramwell and his protégé in a less comfortable frame of mind than was theirs prior to my appearance on the scene'.[47]

He lived at Haslemere in poor health, suffering from his old problems of insomnia and indigestion, until 4 December 1893, when he died after his wife had given him an accidental overdose of chloral. Crichton Browne, now the Treasurer, and Rayleigh represented the RI at his funeral. The Managers promptly arranged a commemorative meeting at which Crichton Browne gave the address, and friends and colleagues added brief words of praise. The last of these was Dewar who spoke of Tyndall as a friend and colleague. He went out of his way to sympathise with Louisa Tyndall and to praise her nursing of her husband in his last years, ending with the strong words: 'She is the noblest woman I have ever known'.[48] A more formal assessment of Tyndall's work was given at the RI three months later by Lord Rayleigh.[49] In 1908 there was a cordial exchange of letters between Dewar and Louisa Tyndall. She had taken objection to a somewhat critical account of her husband's life and accomplishments that Oliver Lodge had written for the 10th edition of *Encyclopaedia Britannica.* Dewar wrote to her on 28 September 1908 saying that he supported her position. He made two points; an encyclopaedia was to record facts and neither praise nor hostile criticism had any place there, and, moreover, Lodge was not a scientist of Tyndall's calibre. The first point is sound but the second more debatable and, indeed, irrelevant. He invited her to lunch, and when she accepted by letter, addressing him as 'Dear Sir James', he asked her to 'Please leave me with the old title of professor'.[50]

Dewar had already embarked on his massive research programme on the liquefaction of gases when he became Superintendent of the House and could, subject to the consent of the Managers, now control all the resources of space and money that were available. Whereas Faraday had regarded himself as a servant of the Institution, his successors had increasingly taken the lead in management, so the forceful Dewar, supported by Bramwell and faced with a generally elderly and complacent body of Managers, could now run things as he wished. He established a strict regime in which there was no familiarity between himself and the staff. On appointment, each of these signed a sheet of duties and conduct, and when new breaches of the conditions of service arose the sheet was revised and each member of the staff had to sign anew.[51] Money was, however, never fully sufficient, a position that all directors of research groups accept as a fact of life. Since 1863 there had been nominally been a Donation Fund which Dewar had urged Tyndall to revive; some equipment (and cash for equipment) was given by Siemens, de la Rue, James Grant (a Manager) and Lachlan Rate (a Visitor) but little was given in the way of free cash until 1884 when Dewar first liquefied air in Britain. From then on gifts of money were frequent and often substantial.[52] Ludwig Mond, who had become a Member in 1883, a Visitor in 1889, and a Manager in 1892, gave £100 each year from 1887 to 1891, when he added a second £100 for the Centenary Celebrations of that year. He then increased the gifts as Dewar's programme of gas liquefaction got into its stride, giving £500 in 1893, and probably about £3500 in all.[52] An unexpected bequest came to the RI early in 1893. Thomas George Hodgkins (1803–1892) was an eccentric confectioner of English birth who lived on Long Island, New York. In October 1892 his lawyer told the Managers that he was making a bequest to the RI of $100000 (then about £20000) for 'an investigation of the relations and co-relations existing between man and his Creator'. This obscure condition was capable of a wide interpretation but, with Hodgkins's known interest in the purity of air, it was decided that the bequest could be stretched to cover Dewar's programme. It was used also to fund two reviews of his work by Agnes Clerke in 1901 and Henry Armstrong in 1908.[53] Two other sources of funding were the Goldsmiths' Company and the Royal Society Government Grant scheme. This had been established in 1850 with the modest sum of £1000 a year and had expanded over the years as scientists became more comfortable with accepting state aid. In all Dewar received £900 in four grants in his earlier years.[54] Later he was denied further support, because, he believed, of the opposition of Michael Foster, the Biological Secretary from 1881 to 1903. Certainly Foster was one of an increasing number of Cambridge scientists who disapproved of what they saw as Dewar's neglect of his duties as Jacksonian Professor. Dewar himself contributed substantially to his research costs after he had succeeded Tyndall as Superintendent – £200 in 1888, £50 in 1889, £50 in

1890 etc.[52] He could now afford such sums since he had two professorial salaries, no children to educate, free accommodation at the RI (although he retained his house in Cambridge) and a substantial if irregular income from his outside work. He was, however, often helpful and generous towards his staff, even going so far as to give Ralph Cory, the Assistant Librarian, a £5 note, a large sum in those days, when Cory congratulated him on his knighthood in 1904.

The laboratories in which he and Rayleigh worked were those created by the improvements introduced by Spottiswoode in 1872, when he was Treasurer, and on which he lectured on 17 January 1873.[55] He had described the old chemistry and physics laboratories respectively as 'makeshift' and as 'a noble relic'. It was, however, not until further works were carried out in 1884, as a consequence of a major failure of the drains,[56] that accurate plans of the building were drawn up. These show a Physics Laboratory of about 110 m^2 on the ground floor behind the Grand Staircase that opened on to Albemarle Street. Below it in the basement was the Chemistry Laboratory of about 165 m^2, with the Engine, Furnace and Battery Rooms, of about the same overall area, to the north of the Chemistry Laboratory. In 1896, at the time when Mond established the Davy Faraday Research Laboratory in 20 Albemarle Street, he acquired also the lease of another 50 m^2 of space in the basement between the two buildings, 21 and 20, which was added to the Chemistry Laboratory and relieved the overcrowding caused by Dewar's massive compressors and liquefaction apparatus.

Dewar's relations with Rayleigh seem to have been similar to those with Tyndall – correct and formal but never warm. The two men met regularly at the RI, although Rayleigh was more often at Terling where he worked in his private laboratory, and so there was little need for written communication. We know that Rayleigh had been disgusted by Dewar's advice on how to manage his relations with his assistants, but there is little evidence that Rayleigh ever took his own into his confidence or looked to them for advice. We know from his son's comment that Dewar was the only man who ever made Rayleigh angry. It was Rayleigh's custom to ask his scientific friends and their wives down to Terling for weekends but it may be significant that the Dewars were asked only once, in 1904, shortly before Rayleigh gave up his Professorship at the RI. Alexander Scott (then unmarried) had been invited three years earlier, although his position was held under the joint supervision of Dewar and Rayleigh.[57] When Rayleigh became Chancellor of Cambridge University in 1908 he conferred honorary doctorates on the President and Secretary of the RI, the Duke of Northumberland and Sir William Crookes, and on George Liveing, about to retire as Professor of Chemistry,[58] but not on Dewar, an omission which was possibly made because he was still in post at Cambridge, or which may have had a more direct implication.

Chapter 7
Commerce

We have seen that, like many of his contemporaries, Dewar held more than one academic post at a time. He was similarly following current practice when he undertook a variety of consulting and advisory posts, and offered his services as an expert witness across a wide range of scientific and technical fields. Faraday and Tyndall had both acted in this way at the RI and there had grown up a practice in dividing the fees between the consultant and the Institution. Dewar's outside work had started when he undertook analyses of water and manures for the Highland and Agricultural Society during his time in Edinburgh. The work on water was to be resumed later in London, and it was from his time there that most of his commercial work arose. In the discussion at the RI over the renewal of his Fullerian Professorship in 1880, at the end of his first three-year tenure, he wrote to de la Rue to say that 'I never intended making any income out of my R.I. chair. The present salary … is sufficient for my work, and if I desire, there are many opportunities of adding to it'.[1] This he soon did.

He acquired the reputation of a reliable analyst who could turn his hand to a wide range of chemical topics and after he left Edinburgh he was a frequent expert witness in the law courts. It is difficult and probably unnecessary to try to track down all the commercial work he undertook. He kept these records privately and his surviving correspondence tells us little of this side of his life. His laboratory notebooks list some experimental results, but in the absence of the supporting reports or letters these are also of little value. Ruhemann was not the only assistant to contribute to his private practice but again Dewar kept all his assistants' papers to himself. Some of his cases are, however, worth study since they contributed not only to the commercial and political life of Britain from, roughly, 1880 to the First World War but the controversies they generated had important consequences for Dewar's professional career and reputation. Other cases bear witness to the range of Dewar's skills.

In an early case in February 1881 he appeared for the plaintiff in a case of patent infringement when one electroplating company complained that another was illegally using its process. The defendant company denied this and said that it made changes which it was unwilling to disclose to the court since they were trade secrets. The court required that their process be shown to Dewar who found that the only change was the addition of a little common salt to the solution and that was without significance for the success of the process. Judgment was given for the plaintiffs. Other cases were to prove more challenging.[2]

A few months later Dewar was the opening witness for the plaintiff in a case that was to have parallels in the more famous 'cordite case' 10 years later. The

German company BASF sued the well-known Manchester firm of Levinstein for breach of an 1878 patent for red dyes. The case went from the Chancery Court in March 1881, to the Court of Appeal in 1885, and to the House of Lords in 1887, the plaintiffs winning in the lowest and highest courts.[3] The case hinged principally on whether the use of the word 'naphthylamine' in the patent was so imprecise as to make the patent invalid. There are two isomeric naphthylamines, the α and the β forms, of which only the former is suitable for use in the process. The α form had been known for many years and was an article of commerce, although not a common one. The β form had been discovered in Germany in 1875–6, and had been described in a few German and English books, but in 1878 it was little known even to laboratory chemists, and quite unknown in commerce. The court held that the patent was valid since the unspecific name 'naphthylamine' could, in 1878, and in a technical context, have meant only the α form. One interesting point is that Dewar carried out his own experiments of synthesis and analysis before giving evidence. The case was heard several months before his appointment of Ruhemann as his assistant and so the work could not have been delegated to him but it was probably just the kind of assignment that Ruhemann was to complain about some years later.

In August 1868 the British Association for the Advancement of Science held its annual meeting in Norwich. Among those attending were three young men who had never before been to such a meeting. The youngest was James Dewar, a few weeks short of his 25th birthday, who read two papers to the Chemistry Section, one on his work on coal-tar bases, and the other on Kekulé's wire-and-ball model of benzene. In the same Section was a paper from a German visitor, the 29-year-old Ludwig Mond,[4] with a more important paper on the recovery of sulphur from the noxious waste of the inefficient Leblanc process for making soda. It was the first major achievement of a career that was to make him a wealthy industrialist. In the Mechanical Science Section, there was a paper from another young visitor, the 34-year-old Alfred Nobel, on his new invention of dynamite.[5] He had patented this in Britain the previous year but his attempts to interest London financiers in exploiting it had no success, in spite of a demonstration of its power and safety in July 1867 in a quarry at Merstham, in Surrey. His BAAS paper also had little immediate effect for the manufacture and carriage of nitroglycerine was banned in Britain for some years from 1869. In the end Nobel found that the financiers of Glasgow were more far-sighted than those in London and he established an explosives factory at Ardeer, on the Clyde, between 1871 and 1873.

It is likely that the two young and sociable chemists, Dewar and Mond, would have met in Norwich, although we cannot now know, but they certainly both met Frankland who was presiding over the Chemistry Section that year and Mond also met Roscoe, from Manchester, with whom he was to have a useful association by his providing Mond's companies with chemists as well-trained as many of those he had brought over from Germany. It is perhaps less likely that either of the two chemists met Nobel. Dewar joined the Association while in Norwich and was to contribute much to its proceedings over the years, but Mond and Nobel did not, and it seems that neither ever again attended a meeting. But Dewar's association with both men was to be long and deep; that with Mond was generally happy and successful, that with Nobel started well but ended in grief.

The three brothers, Robert, Ludvig and Alfred Nobel, involved Dewar in the problems of petroleum through the two elder brothers' role in developing the Baku oilfields in the Caucasus. He later became deeply involved also in Alfred's explosive business. In March 1873 Robert (1829–96) went to Baku from Sweden to buy walnut for the stocks of rifles that the family firm was making for the Russian government.[6] When he arrived he found that there was no walnut to be had there, but that there seemed to be a good opening for investment in the Baku oilfields. He later bought land and a small refinery with the money intended for walnut. From prehistoric times it had been known that oil oozed from the ground on the shores of the Caspian Sea, and that the oil and gas could be ignited, but there had been little in the way of the exploitation of this resource. Ludvig (1831–88) followed his brother to Baku three years later; he was the one with both technical abilities and the managerial skills needed to run the business. He cared also for the conditions of his workers and so acquired a local reputation as a good employer. Robert soon bowed out of active management leaving Ludvig in charge. The third brother, Alfred (1833–96), was meanwhile on the way to making his own fortune from his invention of explosives, of which dynamite, the subject of his BAAS paper in 1868, was the best known.[5, 7] It is formed of nitroglycerine,[8] which is 'tamed' by incorporation into an absorbent powder such as kieselguhr or sawdust. It was a high explosive widely used for blasting, particularly in mining and in the civil engineering works associated in the building of the European railway network. Alfred was also involved in the oil enterprise where he had to persuade Ludwig to moderate his efforts to expand the enterprise more rapidly than Alfred thought prudent.

Dewar's connection with the Nobel family must have begun soon after his appointment to the Fullerian chair at the RI in 1877. On 8 August 1879 Alfred Nobel wrote to Alexander Cuthbert, the Manager of Nobel's Explosive Company Ltd of Glasgow, to say that the letters from Dewar and Frederick Abel were excellent; apparently they had backed Nobel in a dispute over the safety in hot climates of one of his explosives – probably dynamite, but perhaps blasting gelatine or the variety of this called gelignite.[9, 10] Abel[11] was then the Government's leading chemical adviser on explosives through his position at Woolwich and at the gunpowder factory at Waltham Abbey in Essex. He had previously had a dispute with Nobel over the safety of manufacturing and handling of nitroglycerine for use in dynamite, occasioned by Abel's preference for guncotton (nitrocellulose) in which he had an interest that, it was later alleged, may have been financial.[12] This dispute had been resolved by 1871 and relations were cordial again by 1879.

It was clear that Dewar had become involved in the petroleum business by 1882 when he questioned Ludvig on its prospects and Ludvig replied, convincing him that the Russian deposits were too vast to be consumed only in the home market and must be exploited worldwide. In this prediction Ludvig was correct for by 1900 Russian oil comprised over half of the world production, and the Nobel family probably controlled over a quarter of the Russian contribution. The immediate nature of Dewar's advice is not clear but later he was to analyse Baku oil.[13] His fee for his contributions was not a cash payment but the issue to him from 1883 onwards of shares in the family oil company. In November 1884 Dewar was appointed chemist to Nobel's 'company in England', shortly after he had been out to Baku and come back 'full of information about oil wells'.[14] Dewar wanted to reciprocate Alfred Nobel's gifts of shares and other marks of friendship and did so by joining with Abel in nominating him for Fellowship of the Royal Society. The certificate of proposal does not survive but there is a copy in the lists of candidates for 1884 and 1885. Dewar invited Nobel to the Anniversary Dinner on 30 November 1884 so that he could be introduced to influential Fellows and he then compiled a strong list of supporters 'from personal knowledge'; Abel (whose name came first and who was therefore the formal proposer), Dewar, William Thomson, Playfair, Odling, Crookes, Noble,[15] Gilbert,[15] Liveing, Dupré,[16] Tyndall, Frankland, Roscoe, and, 'from general knowledge', the engineers James Thomson, Armstrong,[15] Bramwell[17] and the ex-army officer, Lefroy.[15] Nobel's address was given as 'Ardeer, Scotland and 53 Rue Malakoff, Paris'. He was not a British subject and so was eligible to be proposed as a Foreign Member. But Dewar decided not to do this, and instead proposed him as a Fellow on the strength of his alleged residence in Scotland. It seems likely that when the Council came to look into the proposal they decided that Ardeer was merely the place of one of his several factories; he did not live there and so he was not qualified to be a Fellow, being neither one of the 'natives nor inhabitants of Her Majesty's dominions'. The certificate would then be withdrawn. It is therefore surprising that when Nobel wrote a short account of his career for the University of Uppsala in 1893, when it awarded him an honorary doctorate, he claimed to be 'also a member of the Royal Society (London)'.[17]

Alfred Nobel, having realised Dewar's ability and the wide range of his chemical expertise, tried to involve him in other enterprises, the first of which, late in 1882, was a scheme for the synthesis of ammonia from nitrogen, under pressure, and hydrogen absorbed into palladium where its concentration could be very high. This must have been an attractive idea for Dewar since he had 10 years earlier made experiments on the hydrogen-palladium system under the influence of Thomas Graham's suggestion that the absorbed hydrogen might be there in a metallic state, which Graham had called 'hydrogenium'. Nobel's idea was thermodynamically sound but proved to be impracticable; it was to need much higher gas pressures than Dewar could handle before such a synthesis could be carried out, without the palladium, by Fritz Haber. A second scheme for 'fixing' nitrogen was a reaction between the gas and a mixture of barium oxide and powdered carbon to produce barium cyanide. Here the original idea may have come from Dewar, or from a similar scheme devised by Ludwig Mond, but there seems to be little evidence that Dewar put much effort into the attempted synthesis which was quickly abandoned.[18]

Soon, however, a much more serious matter arose. Nobel's two great contributions to the explosives industry, dynamite and gelignite, were both *high* explosives, that is, their firing led to destructive detonation. Such explosives were used for blasting and as the charge in artillery shells. They could not, however, be used as ballistic or propellent explosives for which a slower deflagration was needed if the charge was not to destroy the barrel of the gun. From time immemorial the explosive used here was 'black powder', or gunpowder, as it was commonly known. It was far from ideal, having insufficient power to impart a high speed to the bullet or shell, and because it revealed to the enemy the position of the firing gun by the black smoke it emitted. The smoke came from the partially unburned charcoal that was left after the deflagration. By the 1870s the search was on for a smokeless powder of greater power but one whose deflagration caused little or no corrosion of the gun barrel. Most of the powers of Continental Europe had made some progress towards this goal[19] and Edward Stanhope, the Secretary of State for War, resolved the Britain must not be left behind – a Committee must be appointed. The chairman was an obvious choice, Frederick Abel, the chemist at the War Office. He asked to be relieved of his formal post in order to take on this new job. The choice of members seems to have been left to Abel. August Dupré had been his colleague on an earlier committee and, no doubt, was quickly chosen. The third member was to be his friend Dewar with whom he had worked on a report on the safety of Nobel's high explosives. Dewar was by then a paid adviser to Nobel's British company, although, to Dewar's chagrin, this appointment ended suddenly with the death of Ludvig Nobel in 1888, but it would probably have been inconsistent with his serving on the Explosives Committee. These names and that of the secretary of the Committee, Captain Jocelyn Thomson, of the Royal Artillery,[20] were approved by Lord Wolseley, the Adjutant-General to the Forces, who queried however the proposed freedom for Abel to make use of his experience as War Office Chemist in exploiting what he might learn from service on the Committee, and who wondered if Dewar was the best man for the post, writing to Abel; 'Would you not rather have younger less well-known men at lower rates of pay?'.[21] (Dewar was then 45.) In the end, Abel, Dewar and Dupré were the three men chosen and they were given a wide remit: to consider 'new explosive agents, or the new applications, or improvements, in the production and application of known explosives which will be referred to the Committee'. One omission in its instructions, which was to give rise to trouble later, was the absence of any mention of its carrying out its own experiments. The Committee was given three years to complete its business. It met for the first time on 17 August 1888 and was soon hard at work, with 11 further meetings before the end of the year. These were usually held in central London, but sometimes at Woolwich, and occasionally at Enfield near Birmingham, the site of the small-arms factory.[22, 23] The business that came before the Committee consisted mainly of proposals for smokeless powders from established companies and from private individuals, both British and foreign. Most of the proposals were based on nitro-compounds, such as nitrocellulose, nitroglycerine and picric acid. Many were from ill-informed individuals and could quickly be dismissed from consideration, but one aroused considerable interest – a German product, based on gunpowder but with less charcoal, that went under the name of Duttenhofer powder, from the name of its inventor. Dewar was in almost continuous contact with Nobel from the autumn of 1888 until the spring of 1889. He hoped the close relations between himself, Abel and Nobel would lead to a smokeless powder that they could recommend to the War Office and he was worried by the competition from Duttenhofer. The position changed radically at the end of 1888 when Nobel came up with the unlikely proposal of a propellent powder based on a mixture of nitroglycerine and nitrocellulose which was not very different in its composition from his high explosive, gelignite. Most experts in the field would have guessed that such a mixture was likely to be too unstable and too violent for a propellent, but he had moderated its violence by incorporating about 6 per cent of camphor, a naturally occurring terpenoid, $C_{10}H_{16}O$, and, he claimed, by changing the 'physical condition' of the components.

It is not clear exactly when this powder came to the attention of the Committee. Their *Final Report*[23] says that Nobel brought, from Paris to London, two samples of a powder 'of an entirely new description' on 14 December 1888, but the more detailed *Proceedings and Reports*[22] describe the receipt of two samples on 1 November, and, two weeks later, Nobel was given a British rifle with which to make further tests in France. The Committee started testing the new powder, now named 'ballistite', on 15 December. The first point at issue was the loss of camphor by evaporation from powder stored at high ambient temperatures. Dewar raised this matter with Nobel on 15 December, a date which argues that ballistite had been received by the Committee before 14 December. Dewar proposed that camphor be replaced by the solid hydrocarbon, anthracene, which was cheaper and, he thought, probably as effective. He and Nobel were practical men and rarely discussed theoretical matters in their long correspondence but in February 1889 he suggested solving the camphor problem in another way by storing the powder in contact with 'a piece or two of solid camphor'. This would have a higher vapour pressure than the camphor in dilute solution in ballistite and so would prevent its evaporation – a 'pretty application of the laws of dissociation' as Dewar put it. Neither resolution of the problem seemed to have appealed to Nobel. He was a shrewd business man but he trusted Abel and Dewar too far, assuming that their testing of his new invention had only the aim of improving it, and that his position was secured by his British patent of 1888.[24] He allowed Captain Thomson to visit his Honfleur factory early in 1889 and see exactly what was being done there. Thomson reported back to the Committee on 30 March to say that the preparation of the powder was still at the laboratory stage and that the composition differed a little from that of the 1888 patent; it was now a 50/50 mixture of nitrocellulose and nitroglycerine, but still with 6 per cent of camphor and a trace of aniline. Benzene was used in the manufacture to give an easily moulded paste but then removed by evaporation from the final product.[25] Meanwhile the correspondence between Dewar and Nobel was becoming less friendly. It is hard to follow its course in detail since all the letters are not now available, since it ranged over a variety of topics and since, as always, Dewar's handwriting is often illegible.

From their first meetings the Committee had agreed that it was impossible to preserve total secrecy and that it might be necessary to take out patents. On 1 November 1888 Dewar had raised the question at the Committee of whether they should now be taking patents in their own names to cover changes in the materials they were testing, but nothing definite seems then to have been decided. This was the day that he probably first saw Nobel's new powder. It was another five months before they went ahead; Abel and Dewar filed a provisional specification of their own first proposals on 2 April 1889, followed by further patents on 27 May and 22 August.[26] Nobel was unhappy with the issue of the April patent and Dewar tried to reassure him the next month by writing that: 'I think that you are labouring under a false impression. The Spec[ification] taken out in the names of Abel and Self are intended for *Government use* and could not have been avoided under the conditions and mode of working of the Explosives Committee.' The changes proposed in the first patent, No. 5614, 1889, were minor and related to the manipulation of the material rather than its composition. The extrusion of the paste through holes to form wires and rods gave rise to the name 'cordite' – apparently a suggestion of Captain Thomson. In the second patent, No. 8718, 1889, they suggested tannin, at 10 per cent, in place of camphor. Tannins are a class of involatile naturally occurring polyphenols. Was Dupré's connection with the wine trade the source of this choice? Another possibility for the moderator was 'vaseline', which was the substance that was eventually adopted by Abel and Dewar, partly under the probably false impression that its lubricating properties might protect the gun barrel from erosion.[27]

Nobel began to be suspicious of where Abel and Dewar's experiments might be leading. On 11 May Dewar tried to reassure him that 'Abel and I still have some magnanimity left'. A meeting on 25 May had led to some kind of understanding. Dewar had met Nobel at his hotel in London at 9 o'clock in the morning, attended a meeting of the Committee at 10, and then he and Abel had met Nobel again at 12. But by July, after some complaints from Captain Thomson of the quality of some of the ballistite they had been receiving, which was not always stable, Nobel wrote to Thomas Johnston, his manager at Nobel's factory at Ardeer, about Dewar's testing of ballistite: 'Dewar hinted to me that it might be used as an argument against the patent, for, said he, if your works cannot make better stuff, surely the patent specifications must be sadly deficient. At Woolwich they invite you to send specimens to establish their inferiority – to theirs. Rocks ahead.'[28] He was right; there were rocks ahead. The relation of Abel, Dewar and Dupré with Nobel is complicated by the parallel discussions on a petition for the renewal of Nobel's patent on gelignite. All three advised the company and as late as 26 February 1890 they were receiving substantial fees from Thomas Johnston; £350 for Dewar, £300 for Abel and £50 for Dupré.

Abel and Dewar's third patent, No. 11664, 1889, introduced a change that was to have profound consequences. The nitration of cellulose produces a range of material whose properties depend on the degree of nitration. The higher the degree the greater the explosive power, so Nobel, conscious that his product might already be dangerously close to the violence of his high explosive gelignite, specified that the nitrocellulose used in ballistite must be of a modest degree of nitration. This degree was conventionally denoted by the solubility of the material. The name of the solvent is often not spelled out but it went under the name of 'ether-alcohol' and was a 50/50 mixture by volume of diethyl ether and ethyl alcohol. In the provisional specification he writes: 'the less of the insoluble nitrocellulose there is mixed up with or contained in the nitrocellulose used … the more sluggish will be the combustion of the explosive material obtained'. This was the result that he was aiming for. The complete specification makes no mention of the insoluble form but is insistent that it is the soluble form that he is using. It contains the phrases 'dry soluble and pulped nitro-cotton', 'soluble nitrated cotton', and even the less correct forms, 'nitrated soluble pulped cotton' and 'nitrated soluble cotton'. Abel and Dewar made the surprising discovery that a high degree of nitration, and so the formation of an insoluble form of nitrocellulose, led to a product that was more stable than the soluble form, but which was no more violent in its deflagration. The relation between solubility and the violence of the explosive is not as simple as they had assumed since cellulose is not a single chemical compound, as was then generally thought, but a complex mixture of molecules of different molecular weights. Nevertheless the solubility and degree of nitration had been taken to be simply linked. Dewar, who may have been responsible for the new proposal, worked on the incorporation of the insoluble form into the paste to be extruded and he succeeded in arriving at an acceptable solution of the problem in July, as he later told the High Court. It was a remarkable and potentially dangerous achievement since common sense told one that the use of the more highly nitrated form would lead to decreased stability and greater violence of explosion, but this proved not to be the case. Dewar was, however, never a man to be deterred by the potential danger of an experiment either to himself or to his assistants. When challenged at a meeting of the Chemical Society by Richard Friswell, Lockyer's assistant and no mean chemist himself, on the risk of a studying the absorption of hydrogen by palladium in steel vessels at 100 atm pressure and 500 °C, Dewar had replied coolly: 'The experiments were dangerous but one had to take the risk.'[29] It was a principle that Dewar followed all his life.

The use of the insoluble form of nitrocellulose was a subject of the third patent of 22 August 1889, which, like its predecessors, was assigned to the War Office. This left undetermined what was to be done about any foreign patents, but here Dewar was in no doubt, telling Nobel that 'With Foreign Patents I may say at once that it never was the intention of either Abel or my self to do anything than to place the same in our *own hands*'. The first were taken out, at their expense, in June 1889 and within a few weeks they had covered eight European countries and the United States. Abel's position seems to have been somewhat uncertain during the life of the Committee. He gave up some of his regular duties at Woolwich and Waltham Abbey after the Committee had been established, but it was not until 10 July 1890 that he told the War Office that he wanted to resign from his regular post as soon as possible, having been wearied by aspersions on his conduct. It is not clear if these 'aspersions' arose from criticism of his part in taking out the foreign cordite patents or from an unrelated accusation of his having overpaid for the nitric acid that his department was buying in bulk. An investigation soon cleared him of this last charge. He assured his employers that he 'would do all in my power, before quitting the Service, to assist in the organisation of the production of the new powder'. He sent in his resignation on 15 November 1890 but it was another nine months before it was accepted by the War Office.[30]

Meanwhile matters moved rapidly in July and August 1889. In a patent of 22 July 1889, Abel and William Anderson,[31] a mechanical engineer in private commercial practice, took out a patent for the machinery to make cordite, and 10 days later Anderson was appointed Director General of Ordnance factories, a position inconsistent with his holding this patent with Abel! He was required to divest himself of any interest in the exploitation of the patent, which he did, but only in form, by transferring his shares in his company, Easton and Anderson, to his wife and his eldest son, Edward, who continued to be in charge of making the new machinery. By the end of July Dewar was satisfied that he could make a good explosive with the insoluble nitrocellulose. *The Times* reporter was later to suggest that his hitting on the use of the insoluble form was arrived at 'partly by chance'.[32] On 5 August the intentions of Abel and Dewar were becoming clear and the latter wrote to Nobel to say that 'No doubt we are *good friends*. There is room in the World for more than one Smokeless Powder'. Four days later, perhaps after he and Abel had recognised the potential of their new discovery, his tone had changed and he wrote that the 'agreement of 25th May is now *abrogated*, so the matter is ended'. Two weeks later they submitted the Preliminary Specification of Patent No. 11664. Abel gave a Friday Evening Discourse on 'Smokeless Explosives' to the RI on 31 January 1890. He mentioned Nobel's ballistite, but not by name, and said that the Explosives Committee was trying to cure its instability by modifications, but made no mention of the use of non-soluble nitrocellulose.[33] He said no more in his Presidential Address to the the BAAS at Leeds that year, but he exhibited cordite at the Jubilee Meeting of the Chemical Society on 24–25 February 1891.[33]

Where in all of this was Dupré? The minutes of the Explosives Committee show that he was actively involved in the testing of the various powders that came before the Committee but apparently not in the experiments that led to the development of cordite. It seems that he had doubts about the correctness of Abel and Dewar's conduct, and his health gave way in 1891 when he had a breakdown of some kind.[16] Dewar clearly considered him a poor colleague, writing later to Abel during the legal actions between Nobel and the War Office 'I sent Dupré away in a very bad state of mind. He seems quite useless'.[34]

By the summer of 1890 it was clear that all powders other than ballistite and cordite had been eliminated from consideration and when the Committee was wound up in June 1891[35] it recommended cordite as the propellent explosive of choice for both artillery and small arms. It was, however, not until 28 May 1893 that this choice was formally ratified by the War Office.

The development and patenting of cordite by Abel and Dewar, one of whom was or had been a civil servant, and both of whom had used state facilities in conducting their research, had meanwhile aroused public attention. On 25 March, Edward Stanhope, the Secretary for War who had set up the Explosives Committee, told the House of Commons as little as he thought he could get away with, saying that 'We understand that cordite powder was made practically by Sir Frederick Abel and another gentleman, and that they immediately assigned the patent to the Government of this country'. The House was not satisfied, and, three weeks later, the Conservative MP for Peckham, Arthur Baumann, asked the Secretary of War 'whether cordite powder has been patented in any foreign country and, if so, by whom?'. The Secretary did not reply in person but put up his Financial Secretary, St John Broderick, to say that 'Patents in connection with cordite have been taken out abroad by Sir Frederick Abel and James Dewar, the inventors'. When he was asked where the profits would go, he was unable to answer.[36] There, for the moment, the matter rested since the Conservative Government lost the election in August 1892 and Arthur Baumann was one of the members who lost his seat. His place in the questioning was now taken by another Conservative MP, Robert Hanbury, the member for Preston, who was a more persistent and better informed critic than Baumann. In August 1893 he asked whether Abel had licensed any manufacturer abroad to make cordite. When this question led nowhere he returned with a full attack on what he called the 'cordite scandal' in the debate on the Army Estimates in September.[37] He had two principal points of criticism, the first of which was that the Explosives Committee had improperly gone beyond its terms of reference by engaging in research and manufacture and not confining its work to assessing the merits and defects of the explosives brought to its attention. How, he asked, could companies or individual inventors be expected to bring their ideas before the Committee if they knew that the Committee could use them as a basis for inventions of their own, with no reliance for recompense for the original inventor beyond what the War Office might choose to make? The alleged competition between cordite and its rivals was, he said, 'a sham'. Thomas Cochrane, the Conservative Member for Ayrshire North, in whose constituency Ardeer lay, said that the Committee had gone further than merely undermining the value of Nobel's invention but had actively hindered his efforts by advertising in the *Glasgow Herald* for workmen, by implication from Ardeer, skilled in making and handling nitroglycerine to come to work at Waltham Abbey.

Hanbury had good ground for his complaint but there was the natural reply that the Committee could not be expected to sit on its hands and do nothing if it received inadequate proposals which further work at Woolwich might turn into useful products. The Committee had recognised this dilemma early in its own deliberations but without suggesting a solution.[38] Hanbury's second point of criticism was the apparent ability of civil servants to profit on their own account from patents taken out abroad that were founded on work in official laboratories, and which might even be of aid to countries that could become Britain's enemies. He contrasted the behaviour of Abel and Dewar with the restraint of Dupré who had 'washed his hands of the whole transaction' and had 'declined to soil his fingers by taking out patents'. Hanbury was well-briefed and had argued his case cogently but he did not persuade the House. Members of Parliament took a greater interest in the technical details of Britain's ability to fight a war than is customary nowadays. They were suspicious of the control of any essential resources such as ballistite or cordite by a foreign company or power and probably thought that Abel and Dewar's ends had justified their means. William Anderson had listened to the debate and wrote the next day to Abel to say that 'There could be no doubt about the feelings of the House. Hanbury did not even venture to take a division'.[39]

Nobel was in a dilemma. Ballistite was a promising propellent but it seemed that it could be trumped by cordite. He was dependent on the good will of Abel if he was to sell ballistite in the important market of Britain and its empire, but Abel was now backing his own powder. Nobel's staff at Ardeer spent many months compiling the necessary evidence of his priority of invention. His patience ran out late in 1893 and he sued the War Office for infringement of his British patent of 1888. The case came before Mr Justice Romer in the Chancery Division on 29 January 1894. Sir Robert Romer had been a notable mathematician in his youth – Senior Wrangler and Smith's Prizeman at Cambridge – and briefly Professor of Mathematics at Queen's College, Cork. The Cambridge Mathematical Tripos, and the questions asked of candidates for Smith's Prize, included 'mixed mathematics', which covered much of what we now call classical physics. Although no chemist, he was therefore probably as well placed as any Chancery Judge to tackle the technicalities of Nobel's suit. The Crown could not be sued in its own court so the case was brought as *Nobel's Explosive Company (Limited)* v. *Anderson*, with the understanding that the Government would stand behind Anderson if the case went against them. It lasted 14 days and aroused great interest. The importance attached to the outcome by the Government was shown by the appearance of the Attorney-General, Sir Charles Russell, to lead for the defence, and the fact that *The Times* reported each day's proceedings in full.[40] The barristers for the plaintiff were led by Fletcher Moulton, another Senior Wrangler, and a Fellow of the Royal Society.[41] He was scientifically a match for any expert witness; one judge had complained of his advocacy that 'Moulton generally gave scientific evidence whilst witnesses argued the law'. Many cases depended critically on what was known to a competent scientist at the time of publication of the patent, not at the time of its alleged infringement, and this knowledge was a matter of fact not law. This was not Dewar's first experience of being cross-examined by Moulton, whom he knew well, for in 1885 Dewar had appeared for the German engineer Nikolaus Otto in a case of patent infringement on the explosive gas mixtures in an internal combustion engine. Dewar had conducted his own experiments and these helped convince the court that Otto was in the right.[42]

Nobel, Abel and Dewar all gave evidence in the 'cordite' case, as did many explosive chemists from Britain and Germany, including Carl Duttenhofer. They were supported by the usual ranks of well-known British chemists, some of whom had made experiments of their own. Odling and Roscoe appeared for the plaintiff and Frankland, Henry Armstrong, Crookes and Bramwell for the defendant. The plaintiff's case rested on the claim that Nobel's patent of 1888 was a 'master patent' from which the work of Abel and Dewar derived and which covered all forms of nitrocellulose, whatever its solubility. It was therefore perhaps unwise for the plaintiff's counsel to venture on a mild joke that drew attention to the distinction between the two kinds. Dewar's letter to Nobel in which he mentioned his and Abel's 'magnanimity' was raised in Dewar's cross-examination. He was asked how much magnanimity he had left in 1889. He replied that he did not remember but he knew how much he had now – 'of the well-known soluble kind, I suppose' replied counsel, to the amusement of the court.

The defence claimed that the 1888 patent was not a master patent but was itself derived from Nobel's earlier 1875 patent on gelignite. They could not deny cordite's parentage – it was clearly a descendent of ballistite – but it was, they claimed, a distinct material. In the minutes of the Committee was the admission that cordite was a 'modification of ballistite devised by the Committee'[43] but it seems unlikely that their use of this phrase was disclosed to the plaintiffs. Nobel had emphasised that he used soluble nitrocellulose while Abel and Dewar had insisted on using the insoluble form, and claimed, crucially, that both forms were recognised as different substances in 1888. Dewar's own evidence on this point was clear. Crookes's first biographer wrote that it was Crookes's words that were of most value in making the defendant's case, but there seems to be no evidence for this claim.[44] It was on the distinction between the soluble and insoluble forms of nitrocellulose that the Judge laid most emphasis in declaring that Abel and Dewar's invention did not infringe Nobel's 1888 patent. He said of Nobel's patent that 'any of the persons to whom this specification was addressed would, I think, from reading it with the then state of knowledge, be led to the conclusion that to carry out the patentee's intention he was to avoid the use of the insoluble form'. He added that Nobel had 'in effect put aside the insoluble form as being outside his intention'. And of Abel and Dewar's patent, he said 'I am not concerned directly with the question of the validity of that patent, but I think it right to say that in my opinion the patentees have arrived at the discovery of cordite only after a long period of experiment and research. In fact they solved the problem which Mr Nobel left unsolved – that is to say, how to make a good powder out of insoluble nitrocellulose and nitroglycerine'.[45]

Nobel was not satisfied by this verdict, not only because of the financial loss of the ballistite market in Britain and its empire, and probably in other countries also, but because of its reflection on his standing as a chemist and engineer. He went therefore to the Court of Appeal, where his case was heard in July 1894 by Lord Esher, the Master of the Rolls and Lord Justices E.E. Kay and A.L. Smith. Again there were expert witnesses for and against Nobel, but only for four days this time. Esher was moved to comment on the problems of such technical cases: 'All chemical experts are like this. My impression is that they come at last to tell you that everything is like everything else, but at the same time everything is different from everything else'. It was a fair summary of much of the evidence in patent cases with which the leading chemists of the day sought to convince technically unqualified judges of the merits of their client's case. Edward Kay had had an encounter with Dewar six years previously in the case *Edison and Swan Electric Company (Limited)* v. *Holland and others*. This was a complicated case that lasted for 21 days concerning the infringement of the plaintiff's patents for making carbon filaments for incandescent electric lights. Kay had thought that Dewar had deceived the court by claiming falsely that he could make such filaments in a brass box, a feat that Crookes was unable to repeat. Dewar was vindicated by further experiments carried out under the supervision of George Stokes and Kay's judgment was overturned in part, on appeal.[46] Nevertheless he now supported Esher's judgment that Nobel's appeal must be dismissed although making it clear that his sympathy lay with Nobel: 'It is quite obvious that a dwarf who has been allowed to climb on the back of the giant can see further than the giant himself.' Here Nobel would have been well-advised to abandon his suit but he was too enraged to see the wisdom of this course and obtained leave to go to the supreme court, the House of Lords. In February 1895 the Lord Chancellor, Lord Herschall and four other Lords of Appeal, were unanimous in dismissing the case, with costs against the plaintiff.[47] It was estimated that the three rounds in the courts had cost Nobel £28000. He inveighed against the English courts, writing a piece of sarcastic drama, *The Patent Bacillus*, and saying 'The moral of the whole cordite case could have been pre-expressed by Hamlet if he had said that there is something rotten in the state of Justice'. He clearly thought that the Government had leant on the courts to give rulings in favour of Abel and Dewar, and was probably strengthened in this view when Abel, who had been knighted in 1883, was 'promoted' to a superior knighthood (KCB, Knight Commander of the Order of the Bath) in 1891, and to a baronetcy as soon as the matter was settled in Chancery in 1893. In May 1894 he became a Vice-President of the Royal Institution. Anderson received a CB (Companion of the Order of the Bath) in 1895. Dewar, however, received no honour then, but he was knighted 10 years later. His brother-in-law, Hamilton Dickson, wrote in his obituary that he had turned down a knighthood some time before 1904, but gave no particulars.[48] It is possible that it was on the occasion of this success in the law courts that the first offer of a knighthood was made and that Dewar thought it wise to decline because of the controversy over his and Abel's behaviour, but this can only be a speculation. Popular opinion had not always matched that apparently shown by the House of Commons, but had proved to be more sympathetic to Nobel's case. So many letters had been written to the press to support Nobel that, after the judgment in the Chancery Court, his solicitors had written to *The Times* to make it clear that they were not behind what looked a campaign to influence the court.[49] In the end Nobel did receive a small *ex gratia* royalty on cordite made for military use at Ardeer. As so often with legal cases, neither side came out of the affair with enhanced reputations. The majority in the House of Commons, and *The Times*, were glad to see Nobel lose his fight for what the paper called 'the maintenance of a gigantic monopoly', a monopoly that could have been to Britain's disadvantage in a war. The public saw two respected scientists indulge for private profit in what seemed to be underhand if legal breaches of confidence. The battle had apparently damaged Nobel's health and he survived only another year, dying at San Remo on 10 December 1896, at the age of 63. The first account of his career was given by H. de Mosenthal, a senior member of the staff at Ardeer for 25 years, in a paper[50] given at a meeting of the Society of Chemical Industry on 1 May 1899. In the discussion that followed the paper, Dupré said that 'he had always publicly and privately expressed his sympathy with Nobel in that case, and thought that he had been somewhat harshly dealt with, though from a legal point of view, perfectly correctly'. Another man who supported Nobel in the same discussion was a junior barrister, John William Gordon,[51] who said that 'He first met Mr Nobel when, to his regret, he had to appear for his opponents in the celebrated case'. Gordon was to become a thorn in the side of Dewar and, indeed, of the RI itself, after his election as a Member in February 1899.

Nobel's death did not quite mark the end of the effect of the 'cordite scandal' on British public life. William Gladstone had led the Liberal Party in its electoral victory of 1892 but his health was failing and he resigned 18 months later, in March 1894. He was then 84. The new Liberal Prime Minister was Lord Rosebery, but he could not form a strong Government and he was defeated in June 1895, on a 'snap' vote in the House of Commons on the adequacy of the country's stocks of cordite. He and his Government resigned and lost the subsequent election. Lord Salisbury formed a Conservative Government in which Robert Hanbury, that 'vigilant and unsparing critic of the estimates' became the Financial Secretary of the Treasury – the poacher had turned gamekeeper. By 1899, three years after Nobel's death, relations between his company and the RI had improved and they gave 100 guineas to the Experimental Research Fund.[52]

The outbreak of the second Boer War in 1899 revealed that all was not well with the high explosives used by British artillery and by naval vessels. Too many shells failed to explode. Another Explosives Committee was set up in 1901, this time under the nominal chairmanship of Lord Rayleigh. He was Dewar's colleague at the RI but Dewar did not become a member of the new Committee, possibly because he did not wish to serve, or possibly because of the ill-feeling generated by the cordite affair. Abel was now too old to be considered; he died the next year at 75. Anderson had died in 1898. Rayleigh did little but hold the ropes and the work of the Committee was done by the other members, Crookes, Noble (now managing director of William Armstrong's munitions business) and Roberts-Austen of the Royal Mint. They found that as well as the problem of failing shells there was a problem with the cordite then in use which was eroding the barrels of large guns. They remedied this by reducing the proportion of nitroglycerine used in such guns from 58 per cent to 30 per cent. There was no problem with small arms for which the original cordite continued in use.

There was a core of well-established British chemists who appeared time and time again in the patent cases of the 1880s and 1890s; Roscoe, Frankland, Odling, Crookes, Henry Armstrong and Dewar were perhaps the most prominent. Of these all but Crookes held nominally full-time appointments and their earnings from a case in the Chancery Court were generally about £30 to £50, although more had been paid in the 'cordite' case. These sums were a welcome but not an essential addition to their incomes. Crookes was self-employed and he had for some years complained that his and his colleagues' work as expert witnesses was underpaid. A breakthrough came in 1894 in a case involving the extraction of gold from its ores by means of a solution of potassium cyanide. The chemistry was confused and the case, which went to appeal, turned on the precise wording of the plaintiff's patent; the appeal court found it defective. Roscoe, Crookes, Roberts-Austen and Dewar all appeared, and Crookes and Dewar argued successfully for greatly increased fees; they received 150 guineas (£157.50) each.[53]

Soon after disentangling himself from the problems of explosives Dewar was involved again in the analysis of drinking water. In 1882 the London Water Boards had set up a committee to make regular analyses of what they were selling to their consumers. The three men who undertook this job were William Crookes, William Odling and Charles Meymott Tidy, who held the comprehensive post of Professor of Chemistry, Medical Jurisprudence and Public Health at the London Hospital Medical School. He was only 39 but was already recognised as the authority on the combined chemical and bacterial analyses of water. Their reports were carried each month in Crookes's journal, *Chemical News*. In March 1892 Tidy died of cancer, leaving Crookes and Odling to struggle on. Their work came under scrutiny that year when the 'Balfour Commission' was set up in March to examine the quantity and quality of the London water supply.[54] The chairman, Lord Balfour of Burleigh, sat in the House of Lords as a representative Scottish Peer and was Secretary of State at the Board of Trade until the Conservatives lost the election in August 1892. Dewar was appointed a member of the Commission, which met first on 16 May 1892 and reported on 8 September 1893. In that time it sat for 45 days, on 32 of which its members examined 92 witnesses. Dewar, thorough as usual, missed only two days of these examinations. The question of the quantity of available water became a matter of the geology of the Thames basin – Sir Archibald Geikie was the geologist on the Commission. The quality was discussed by representatives of the water companies, by Medical Officers of Health and by chemists.[55] The commission was satisfied on the grounds of the quantity and of the chemical purity of the water but had problems with the bacteriological issues. In its Report the Commission confessed that 'Bacteriology, however, is still in its infancy'. They discussed whether malaria, cholera, typhoid and enteric fever were diseases that could be carried by the water supply. The chemists did their best in this field, helped by Dewar who tried to bring some order to the their evidence, but in the end, the Commission had to leave open some of the questions of the adequacy of the testing of water for organic carbon and nitrogen, their origin and sanitary significance, and, as is usual for committees, they recommended further work. They noted, however, that there were now relatively few cases of possible water-borne diseases.

By July 1894 Odling, who lived in Oxford, had had enough of the onerous duty of analysing water for the London companies and resigned from the committee. Crookes was not sorry to see him go since he was not pulling his weight. The next month Crookes recommended the indefatigable Dewar to the companies and so he took over Odling's well-paid post. Crookes and Dewar set up a laboratory in a private house in Colville Road, in Notting Hill, conveniently close to William Crookes's home. This was managed for them by Crookes's eldest son, Henry, who also lived nearby. This arrangement survived until 1905 when the number of analyses being demanded outran the resources of a private laboratory and in May the newly formed Metropolitan Water Board asked Crookes and Dewar for their advice. This was that the Board should set up its own laboratory which it did in October, with full-time staff. However Crookes and Dewar then established their own testing laboratory, the Water Supply Company, which employed Henry who took up the bacteriological side of the business. This extension was to lead to a row between Dewar and the Crookes, father and son, which spread beyond the question of water and is discussed below (Chapter 11).

Ludwig Mond's improvement to the operation of the Leblanc soda process, reported to the BAAS in 1868, was soon overtaken by Ernest Solvay's invention in Belgium of the more efficient ammonia-soda process. Mond at once recognised the potential superiority of Solvay's process and also of the practical difficulties of working it on an industrial scale. He obtained a licence to exploit it in Britain and embarked on a deep study of its strengths and weaknesses. It was his improved process that was the foundation of the company of Brunner and Mond, which was established in 1873. When Dewar was President of Section B, Chemistry, at the meeting of the British Association in Sheffield in 1879, he spoke on this important new chemical process.[56]

His industrial career now secure, Mond was able to return to the more fundamental research that was his real love. In 1884 he came up to London from his provincial works and sought there a home where he could meet the scientific elite in a way that he could not in Cheshire. He soon had a large house, The Poplars, in Regent's Park, where he and his family could live and entertain in style, and where there was room for him to set up his own private laboratory. He became a Visitor at the Royal Institution in 1889, and a Manager in 1892. He contributed £100 a year to its research fund for some years from 1887, increasing the gift to £500 as the scale of Dewar's work on the liquefaction of gases became apparent (see Chapter 8).[57] In return for Mond's support, Dewar sought his election to the Royal Society, as he had done previously for Nobel. This time, however, he was on more solid ground for Mond certainly lived in Britain and he had become a British subject in 1880. The election certificate, dated 27 February 1889 and in Dewar's hand, was signed by 15 Fellows from their knowledge of Mond's achievements, and three more from 'general knowledge'.[58] Many of these 18 Fellows had also signed Nobel's certificate: Crookes, Gilbert, Abel, Dupré (now clearly back in favour), Odling, Liveing, Playfair and, from general knowledge, Ramsay. The new men were Thorpe, Roberts-Austen, Hugo Müller,[59] George

Matthey,[59] Raphael Meldola,[59] Warren de la Rue, and, from general knowledge, Vernon Harcourt[59] and William Hartley. Mond had to wait three years but was elected a Fellow on 4 June 1891. On 16 November 1893 he was Dewar's guest at the 150th anniversary dinner of the Royal Society Club, a venerable dining club that is still in existence.[60]

Dewar never became as involved in Mond's concerns as deeply as he had been in those of the Nobel family, but he did give evidence in one case concerning what came to be called Mond Gas. If steam is passed over hot coal then it partially oxidises the coal and the gas that is formed is a mixture mainly of hydrogen, carbon monoxide and nitrogen. It is an excellent clean gas of high calorific value, but unfortunately the reaction by which it is formed is endothermic, that is, it requires a continuous input of external heat if it is to be operated continuously. If, however, the hot coal is partially burnt to carbon monoxide by a stream of air then the gas formed, producer gas, is generated exothermally, but the calorific value is much reduced. Mond blended the two processes by using a stream of a mixture of air and steam, so as to balance the overall reaction thermally and to produce a gas with a good calorific value which could be used for industrial heating. Moreover the ammonia content of the original coal could be recovered and it was this valuable product that Mond needed for the ammonia-soda process. The gas was usually called Mond Gas and the process was exploited by the Power Gas Corporation. To sell the gas he sought to create a mains system in the industrial Midlands from which any firm could draw the gas as needed. The South Staffordshire Mond Gas Company was incorporated on 3 December 1901, with a local magnate, Sir Benjamin Hingley, as chairman, with Ludwig Mond as deputy-chairman, and its headquarters at Tipton. This novel proposal at once ran into opposition from the town gas suppliers who saw a threat to their own mains system. Dewar and other eminent scientists, including Lord Kelvin and Frederick Bramwell, supported Mond, who won the case. By 1923 there were 160 firms in the Midlands drawing their gas from Mond's mains.[61]

Dewar was not involved in Mond's greatest discovery, nickel carbonyl, in 1890, and its use as the basis of the industry for extracting and purifying nickel,[62] but he was clearly interested in the unusual properties and reactions of this and other carbonyls. He and Liveing studied the spectrum of its flame and allowed Mond to show it in a Friday Evening Discourse to the RI before their formal publication. Ten years after Mond's discovery Dewar embarked on a series of papers based on work done at Cambridge by his Demonstrator there, H.O. Jones; their content is sufficiently summarised by their titles.[63]

Another important case in which Dewar played a part as an expert witness was that of the attempt in May 1905 by the Baeyer Company to defend its British patent for the manufacture of aspirin. This substance, acetyl salicylic acid, had had a complicated history; it was first prepared by the French chemist Charles Gerhardt in 1853, but he did not establish the structural formula. The German chemist Karl-Johann Kraut, at the Technical University of Hanover, made the substance in 1869 and was the first to set out its structural formula, but he never claimed to have prepared it in a pure form. In 1898 the Farbenfabriken vormals Friedrich Baeyer and Co. took out a British patent in which they claimed that 'on heating salicylic acid with acetic anhydride, a body is obtained, the properties of which are entirely different from those described by Kraut'.[64] Another German company, Chemische Fabrik von Heyden, was importing aspirin into Britain and so was sued by Baeyer for the infringement of their patent. The case came before Mr Justice Joyce in the Chancery Court on 2 May 1905. As in the 'cordite' case, the eminent patent lawyer Fletcher Moulton led for the plaintiff. He was supported this time by Dewar and by Adolf Liebmann (1852–1927), a German chemist who had worked with Kekulé, who came to Britain to work first in Ivan Levinstein's dyestuffs factory in Manchester, then came to London in 1901, and was to partner Dewar in several patent cases.[65] They were opposed by Percy Frankland[66] and Dewar's friend, Henry Armstrong. As in the 'cordite' case, Moulton could not carry the day for the plaintiffs. The judge roundly condemned the claims in the patent:

> It would be a strange and marvellous thing ... [if] an ingenious person, by merely putting forward a different, if you like, a better mode of purification from that stated by Kraut to be feasible, could successfully claim as his invention and obtain a valid patent of use as a new body or compound. In my opinion it was not a new body or compound, and I hold the patent in question in this case to be invalid ... this specification contains no element of invention beyond what was common knowledge.[67]

Baeyer was left only with the undisputed right to the word 'aspirin' as a trade mark.

Dewar was again paired with Moulton in the defence of Welsbach's patent for incandescent gas mantles based on the oxides of the 'rare earth' metals. This time both were on the winning side.[68]

Towards the end of his career Dewar became involved in a case that concerned coal-tar derivatives, the field of his early chemical research. Vidal Dyes Syndicate had patented a process for the manufacture of a red or brown dye which they claimed had been improperly copied by Ivan Levinstein and by another company, Reed Holliday and Sons.[69] The details were complicated. Moulton became involved as a Special Examiner to advise the court, the plaintiff's evidence being put before him by Horatio Ballantyne, a young Scots chemist employed by Adolf Liebmann, who also appeared later at the trial partnered, as on earlier occasions, by Dewar. The plaintiffs won but Ballantyne found working with Dewar a far from pleasant experience (see Chapter 10).

Dewar did not always patent his own discoveries but, as Henry Savage Landor put it, possibly with some exaggeration, 'if friends of his took advantage of them [his inventions] he became involved in worrying law-suits which lasted years, on technical points, cost him thousands and ended in nothing'.[70] Nevertheless there are at least 25 patents in his name covering a period of 40 years. Eight of these, in the field of explosives, were an important group, but the rest are mainly minor improvements to processes or equipment and are of little consequence. There was, however, one omission that must have lost Dewar a lot of money, namely his failure to patent the vacuum flask that he relied on in his work on liquefying gases.

Chapter 8
Cryogenics[1]

8.1 Early Experiments

It is hard to know when Dewar first resolved to make the study of matter at low temperatures his principal field of research. He had spoken on the latent heat of liquefied gases at the meeting of the British Association in 1874, but his paper was not printed so we do not know what he said.[2] He was a keen student of the history of chemistry, going so far one year as to choose alchemy as the subject of the customary course of six lectures 'for a juvenile auditory' that was offered, after each Christmas, in the Royal Institution. When he accepted the Fullerian chair there in April 1877 he knew much of Davy and Faraday's work on the liquefaction of such gases as chlorine, ammonia, nitrous oxide and hydrogen sulphide. Faraday had been the first to liquefy six such gases. He and his contemporaries had realised that although many gases could be liquefied by pressure at temperatures accessible with solid carbon dioxide, others required cooling to temperatures well below this before liquefaction could be induced by compression, however great. The distinction between the two classes was finally made clear by Thomas Andrews in experiments at Belfast in the 1860s and set out in his Bakerian Lecture[3] to the Royal Society in 1869. For each substance there is a critical temperature above which the liquid state, in equilibrium with its vapour, cannot exist (Figure A.1).[4] If, for any substance, this critical temperature is above about 200 °C then that substance is usually already a liquid at atmospheric temperature and pressure. If it is lower, then the substance is usually a gas but it may be readily condensable. Thus ammonia, with a critical temperature of 133 °C, is condensed at room temperature (say, 20 °C) by a pressure of seven atmospheres or more. Carbon dioxide, with a critical temperature of 31 °C, condenses at room temperature only if the pressure is above about 50 atmospheres. Ethylene (T_c = 9 °C) cannot be liquefied by pressure alone at room temperature but a modest degree of cooling to the ice-point at 0 °C allows the gas to be liquefied by a pressure of 40 atmospheres. There were, however, what were called informally the 'permanent gases', such as hydrogen, nitrogen and oxygen (and so air also), for which the necessary degree of preliminary cooling was large but still unknown. This interplay of pressure and temperature is the recurrent theme of this chapter.

Heating a gas is a simple operation and compressing it is almost as easy, but cooling a gas requires a means of abstracting energy from it, and that is a much harder problem. It was one that attracted attention in the early and middle years of the nineteenth century from engineers who needed cold conditions for the brewing industry, particularly in Germany, and for the preservation of meat on its transport to Europe from the Antipodes. The temperatures needed were never below about -20 °C and so the emphasis of the necessary experiments was on thermal and mechanical efficiency rather than on achieving the lowest temperatures.[5] There was therefore a high level of engineering knowledge in this field which served the physicists and chemists well when they tackled the more difficult problem of attaining extremely low temperatures.

The energy of a gas has three parts; the first, and usually the most important part, arises from the kinetic energy of the motion of its molecules. It is this energy that measures the temperature, for the absolute or thermodynamic temperature is proportional to the average of the square of the speeds of the molecules. The second part arises from the internal energy of the rotation and vibration of each molecule, which also changes with the temperature. The zero from which the internal energy is measured is arbitrary, for we are concerned only with changes of this energy. The third part comes from the energy of interaction of each molecule with its neighbours, which is often small and is the most difficult to measure or to estimate. In a gas at equilibrium there is a free interchange of energy between these three components. A hypothetical gas in which the third, or intermolecular energy, is zero for all positions of the molecules is called a perfect or ideal gas; it has particularly simple mechanical and thermal properties. Real gases approach this limit of perfection as their pressure is reduced towards zero and so the average separation of the molecules is increased. There are therefore two ways of abstracting energy and so cooling the gas, first by lowering the kinetic energy directly by making the gas do work, and second by changing the state of the gas in such a way that the energy of the attractive intermolecular forces is reduced. Since the first energy is the more directly accessible, the earliest attempts at liquefying the 'permanent' gases all used the first route; the second was only introduced later and in such an indirect way that it seems that even those who used it did not always understand what they were doing. Its great advantage, however, was and, indeed, still is, that it is mechanically more simple. For a perfect gas only the first method would lead to any cooling.[6]

Dewar had only been in post at the RI for eight months when the field of gas liquefaction suddenly sprang to life at the Monday meeting of the Académie des Sciences[7] in Paris on 24 December 1877. Two men announced there that they had liquefied oxygen, independently and by slightly different methods. They were Louis Cailletet, a metallurgist in the family iron-works at Chatillon-sur-Seine, in Burgundy, and Raoul Pictet, a physicist in Geneva. Cailletet was at the meeting in person since he had been elected a Corresponding Member of the Academy a week earlier. Pictet's announcement was made by means of a telegram sent two days previously, which was followed at once by a fuller account of his experiment. They had both chosen to work with oxygen since it was the easiest of the 'permanent' gases to prepare in a pure state, and probably because the early indications showed that it would be easier to liquefy than nitrogen or hydrogen. Cailletet had compressed the gas to about 300 atmospheres in a stout glass tube which was cooled by liquid sulphur dioxide, which boils at -10 °C at atmospheric pressure. On opening a valve most of the gas was released into the atmosphere but what was left had been cooled by the abstraction of energy from the gas in the form of the work used by the lesser part in expelling the greater part (Figure A.2). The cooling was sufficient to induce some condensation of the remaining gas in the form of a mist of fine droplets. In his first experiments he had assumed that these were formed of some impurity in his gas but later experiments convinced him that the drops were, indeed, liquid oxygen. They made, however, only a fleeting appearance before evaporating by contact with the walls of the glass tube. He liquefied carbon monoxide in the same way and a week later he was able to report that he could also liquefy nitrogen. He said at first that he had no success with hydrogen, but later[7] was to claim that he had seen a mist also with this gas, but he must have been mistaken with the gas or the experiment since his conditions were later to be shown to be irreconcilable with those needed for the liquefaction of hydrogen. His was not the last of such premature claims.

Pictet had proceeded rather differently.[7, 8] He had also compressed his oxygen to about 320 atmospheres, he then cooled it first by circulation through a tube immersed in liquid sulphur dioxide, and then cooled it further in a second cycle with liquid carbon dioxide to which he had added ethyl ether to form what he called 'Thilorier's mixture'. This addition of ether allowed the temperature to be lowered below that of the gas-liquid-solid triple point of pure carbon dioxide of -67 °C, or 216 K. These two cycles alone were not sufficient to liquefy the oxygen, whose critical temperature is 155 K, but on releasing the pressure, as Cailletet had done, he obtained a jet of gaseous and liquid oxygen. This use of a succession of liquids to achieve increasingly low temperatures was later to be called the 'cascade process'; it was clearly capable of further development in the way that Cailletet's 'one-shot' method was not. Neither man had been able to obtain the liquid in bulk but they had broken the spell; the 'permanent' gases *could* be liquefied. Jules Jamin, in the discussion that followed Cailletet's paper noted, however, that: 'The definitive experiment is still to be done; it will consist in maintaining the liquid oxygen at its boiling point'. Jamin provided space in the Physics Department at the Sorbonne for Cailletet's later experiments. The Royal Society's Davy Medal for a recent discovery in chemistry had been awarded first to Kirchhoff and Bunsen in 1877 for their spectroscopic discoveries, and the second award in 1878 now went to Cailletet and Pictet for their liquefaction of oxygen. The phrasing of the award was, however, likely to mislead casual observers of the extent of their achievement, namely: 'This award is made to these distinguished men for having independently and contemporaneously liquefied the whole of the gases hitherto called permanent.'[9] Dewar, in particular, was always keen to explain the limited degree of their success, as had Jamin at the meeting at which Cailletet's discovery was announced. It is, however, remarkable that the notoriously conservative chemists were so willing to recognise such 'physical' achievements in each of their first two years of the award.

The news of their successes was soon circulated in the scientific world. On 27 December the journal *Nature* reported Pictet's experiment[10] from an account in a

Geneva journal. The printed issue of the *Comptes Rendus* of the Paris Academy followed soon after; in Oxford the Radcliffe Library received that issue of the journal with the two papers and their discussion on 2 January. *Nature* followed the next day, 3 January, with a report of both experiments and a clear description of Pictet's apparatus.[10] The editor, Norman Lockyer, lamented 'Why then in England, the land of practical science, have we not more men like MM. Cailletet and Pictet'.[10] His plea was soon to be answered when Dewar entered onto the scene.[11] Later in January, Cailletet's apparatus was also described in *Nature*[12] and copies of this apparatus were soon on the market. The Paris instrument maker, Eugène Ducretet, had a great stock of physical, electrical and mechanical equipment which he sold mainly for teaching rather than for research. He now made a neater version of Cailletet's apparatus, under instruction from the inventor, which he put on sale; his name is seen clearly on the woodcut published in May[12] in *Nature*. The 'Cailletet pump' is a simple piece of equipment, modelled on Andrews's apparatus of the 1860s, in which the gas under study is compressed in a stout glass tube over mercury, the mercury being itself compressed manually by forcing a piston into it.

There are three essential steps in making useful samples of the liquefied 'permanent' gases. The first had been taken by Cailletet and Pictet when they had demonstrated mists of transient droplets, either in a glass tube (Cailletet) or in a jet ejected through a valve (Pictet). The second step is to find a way of collecting this liquid in sufficiently large amounts for it to form a bulk liquid with a visible meniscus separating it from its vapour. This step was made in Krakow in 1883. The third was to find a way of insulating this liquid so that it did not boil away before it could be studied, and so that it could be used to cool other material and to carry out experiments at low temperatures. This last step was made by Dewar at the RI in 1892 when he invented the vacuum flask that still carries his name today.

Zygmunt Wróblewski was in Paris, studying under Sainte-Claire Deville, when Ducretet put his 'Cailletet pump' on the market. He bought one which he took back to Krakow when he was appointed to the chair of physics in 1882. He found there a kindred spirit in Karol Olszewski in the Chemistry Department who was working in the same field with inadequate resources.[13] They began to work together and by April 1883 they had succeeded in changing the apparatus of Cailletet and Pictet in such a way that they were able to collect liquid oxygen in bulk, although not yet in large quantities. It is probable that Wróblewski was the principal instigator of their experiment; his name is first at the head of the paper, which is marked as coming from the Physics Department. They compressed oxygen and passed it into a stout glass tube cooled with liquid ethylene. This is a better coolant than the carbon dioxide used by Pictet since it remains liquid down to 104 K. It boils, however, at 169 K at atmospheric pressure and this is still not cold enough to liquefy oxygen whose critical temperature is 155 K. Their principal improvement was to cool the ethylene further by evaporating it under a reduced pressure of about 25 mmHg, or 0.04 bar. Here its boiling point is about 130 K. Oxygen can be liquefied at this temperature by a moderate pressure of about 20 atm. In this way they were able to obtain small quantities of liquid oxygen in a glass tube and see that it was a clear colourless liquid; later and larger samples were found to be pale blue. By using their liquid oxygen as a coolant Wróblewski claimed also to have seen a mist of liquid hydrogen, but he and Olszewski often disagreed about hydrogen.[14]

Unfortunately ethylene was expensive and is dangerously flammable or explosive in the presence of liquid oxygen. Dewar was never one to be put off by possible hazards and it became his coolant of choice when he entered the field. By 1892 he had acquired about 'half a hundredweight' (50 kg) for his experiments, but in 1886 he had suffered the inevitable explosion while convincing a visiting lady that he had made liquid oxygen, but had forgotten that there was also liquid ethylene in the laboratory. After the explosion de la Rue was the only man present who could go for a surgeon. Dewar was incapacitated for some months.[15] He told Kamerlingh Onnes in Leiden that the explosion had nearly killed him, and that he had stopped work on liquefaction until 1889.[16] He had, in 1883 at the British Association meeting at Southport, suggested the equally dangerous liquid methane as a coolant, noting that its critical temperature was below 173 K and its critical pressure probably as low as 39 atm.[17] (191 K and 46 atm are now known to be the true values.) Cailletet had made the same proposal in a *pli caché* that he had deposited with the French Academy in 1881.[18] Wróblewski used liquid methane as a coolant three years later.[19] Meanwhile Cailletet also used liquid ethylene but he cooled it not by evacuation but by a stream of cold hydrogen blown over its surface.[20]

As we have seen, Warren de la Rue, a Manager and soon to be the Secretary at the RI, gave the Institution £50 and a 'Cailletet pump' in March 1878. The pump was used first by Dewar when he liquefied acetylene on 11 April at a Thursday afternoon lecture; its vapour pressure at room temperature is about 45 atm. His assistant, Gerrard Ansdell, used the pump for his work on the critical points of fluid mixtures (1879–82), Dewar used it for some mainly qualitative work on the change of the critical temperature and pressure of carbon dioxide on adding other gases (1880), and for measuring the melting-point of ice as a function of pressure (1880).[21] This last experiment was an extension of the classic experiment of William Thomson on the same subject in 1850 in which he had followed the melting curve to 17 atm in order to confirm his brother's theoretical deduction that the temperature of the melting-point of ice, which contracts on melting, would be lowered by increased pressure. The work of the two brothers was a crucial experiment in establishing the fundamentals of thermodynamics. Dewar measured the melting point to 250 atm and made qualitative observations up to 700 atm, but found nothing really new. The Cailletet pump was to be the foundation of Dewar's entry into the cryogenic field, marked by his second Friday Evening Discourse as Fullerian Professor on 14 June 1878 and by an unpublished paper at the BAAS meeting in Dublin that year. In his Discourse he said that de la Rue's gift was 'for the purpose of illustrating this lecture' but clearly Dewar had wider uses in mind.[11] He recounted there the history of gas liquefaction, dwelling mainly on Faraday's work at the RI, and again illustrating his talk with a tube of liquid acetylene. He relied on Cailletet's claim that he had succeeded in liquefying hydrogen also and so he concluded his Discourse with the triumphant claim: 'The work of Faraday has been completed: every gas may be forced to appear as a liquid.' His own career was to bear witness to the falsity of that claim of completion.

After the example set by Wróbewski and Olszewski in 1883 there was no real difficulty in preparing small samples of liquid air and Dewar achieved this publicly for the first time in Britain on 13 June 1884 at a Friday Evening Discourse at the RI. Two weeks later he repeated the lecture for the benefit of the Prince and Princess of Wales, and he also spoke on liquid air at the Montreal meeting of the British Association later that summer.[22] He returned to the subject the next year with a Friday Evening Discourse on 'Liquid air and the zero of absolute temperature', but little is known of that occasion, apart from Archer Hirst's complaints, and the fact that it too was repeated for the Royal couple.[23] Dewar was often reticent in public about the form and construction of his apparatus, saying that he was always introducing changes and so it served no good purpose to describe it at any particular moment.[24] Nevertheless the 1884 Discourse is accompanied by a sketch of a simple apparatus, similar to that developed in Krakow, which showed that he had oxygen in a glass tube at about 150 atm cooled by liquid ethylene pumped down to a pressure of 25 mmHg and so at a temperature of 133 K. Since the critical temperature, 'according to Polish observers', is thought to be at about 140 K (actually, 155 K) the oxygen condenses in the inner tube. Its amount, as could be seen when an image of the tube was projected on to the lecture-room screen, was only about 1.5 cm^3. Whether there were substantial changes before the 1885 Discourse is not clear but it seems likely since Hirst complained about the noise of the pumps used, and the simple apparatus of the 1884 Discourse was unlikely to have given rise to that objection. In 1885 Dewar also showed an apparatus in which he liquefied methane before an audience at the Society of Arts in a lecture on 'The American oil and gas fields'. Here solid carbon dioxide was used to liquefy ethylene at 173 K, which sufficed to liquefy methane (T_c = 191 K).[25]

This work could not have been accomplished without the help of Robert Nicol Lennox whom he had employed as his assistant in December 1881. Lennox had a junior colleague, John William Edward Heath, who was appointed the same year.[26] Dewar was always meticulous in acknowledging the help of these two men in his papers and lectures, with carefully graded praise which made it clear that Heath was the junior. Lennox was a Scot from the village of Killin on Loch Tay in Perthshire. He studied under William Dittmar at Anderson's College in Glasgow and later worked for him on the analysis of sea-water samples from the *Challenger* expedition. Dittmar was a chemist whom Dewar admired and with whom he had worked briefly in Edinburgh; it was presumably this connection that led him to invite Lennox to London. Dittmar was himself not the easiest of men to work with and Lennox's training in Glasgow probably stood him in good stead when he came to work for Dewar. Lennox was primarily a chemist but he claimed to have had some training also in engineering from James Thomson who had taught that subject at Glasgow University. In the Census return for 1891 he was living in Fulham with his wife Ellen, his younger brother Peter, an 'electrical engineer', and two children. He was described as 'Assistant Professor of Chemistry' at the RI, but this was presumably the census-taker's misunderstanding of 'Assistant *to the* Professor of Chemistry'. By the time of the Census of 1901, when he was 40, Lennox was describing himself as a 'chemical engineer', which was not then a common calling. He was engaged at the RI on an annual salary of £100 (with Heath on £60, shortly raised to £80). Over the years Lennox advanced by steps to £180 per annum in 1892, with a supplement paid by Dewar out of his own pocket, for he was generous to those he thought worthy of it. Lennox was by then a Fellow of the Chemical Society, having been proposed by Dewar and supported by a bevy of Dewar's associates, including Liveing, Abel and Ludwig Mond. Dewar was, however, careless over Lennox's address which he gave as 3 Inverness Terrace, Putney, whereas he lived on the other side of the Thames in Fulham.[26] In 1916 he became a member also of the professional body, the Institute of Chemistry. Lennox was seriously injured in June 1896 when a pressure gauge burst and he lost an eye. (The 4th Lord Rayleigh said in 1936 that it was liquid acetylene under pressure.) Ten years later Heath also lost an eye, and Lennox was again injured, in an explosion. By then the RI had accident insurance in place, as required by the Workmen's Compensation Act of 1897. These accidents both followed earlier explosions in 1893 and 1895, Dewar's own accident of 1886, his possibly rash work on cordite in 1889, and there was to be a fire in the Davy Faraday Research Laboratory in 1902, but these events seem not to have diminished his appetite for taking risks.

He was not present, however, when his Demonstrator in Cambridge, William Spivey, died after a laboratory fire in 1901.

The 4th Lord Rayleigh remembered that:

> Lennox admired Dewar and always spoke of him respectfully. Behind the scenes it was very different, and plain speaking on both sides was the order of the day. Indeed I have been in the laboratory while delicate manipulations were in progress and whenever Dewar made a suggestion as to how some particular operation should be carried out, Lennox made it a point of honour to contradict him, and do something different. So far as I could observe this *riposte* did not generate any ill feeling, the parties understanding one another well, and finding it rather stimulating than otherwise. On other occasions it went further, and really hard words passed. … Finally in 1908 the end came. It appears that an expensive plant designed by Lennox for the separation of helium from the Bath gas by rectification failed almost completely. Dewar complained in no measured terms, and Lennox made his reply. If they had limited themselves to words the breach might have been healed: but they started to exchange letters and a permanent quarrel resulted.[26]

It was near his home in Fulham that Lennox had branched out, presumably with Dewar's knowledge and consent, by establishing a small engineering shop which he called the 'Rodebush Works'. The Managers' Minutes of the RI record the first payment of nine guineas (£9.45) to 'Lennox, Reynolds and Co, Electrical Engineers' on 3 July 1893, but the word 'Electrical' was dropped in December 1894.[27] Lennox's partner was Kenneth Reynolds, 10 years his junior, who described himself as 'Secretary of Lennox, Reynolds & Co. Ltd' in the 1901 census. His contribution to the company was presumably financial and administrative rather than technical. Lord Rayleigh said that the financial state of the company was such that it required more than one reconstruction.[26] Kenneth Reynolds was the son of Rev. Henry Reynolds, and brother of Katherine Marian Reynolds (b. 1860) and Sidney Reynolds,[28] both keen photographers. Marian spent much time at the RI photographing Dewar's equipment and experiments in the 1890s. She photographed the lecturer's bench for Dewar's Friday Evening Discourse in 1891 on the occasion of the centenary of Faraday's birth, and, on another occasion, Lennox sitting on the steps of the Lecture Room.[29] Dewar encouraged her in recording the internal aspect of the RI on which he had expended much care and expense. In 1891 she was living with her widowed father in Hanover Square, where he died five years later. In 1895 there was an exhibition of scientific and technical photographs at the Imperial Institute, for which Dewar's friends Armstrong and Abel, the Director of the Institute, were on the organising committee, and where Lord Rayleigh and Marian Reynolds exhibited photographs. (It is the recording of her name here that suggests that she was known as Marian Reynolds, not Katherine, which was also her mother's name.)[30] From March 1897 the engineering firm became Lennox, Reynolds and Fyfe. The Library Assistant from 1890 was Herbert Fyfe, but he had no engineering skills, so was almost certainly not an active partner in the enterprise, but it is possible that either he or a relative put some capital into the firm and so became a sleeping partner. Fyfe left the Library in 1901. There was another man with a similar name, Henry Fyfe, a chemical friend of William Ramsay, but it is unlikely that he was involved in view of the developing antagonism between Dewar and Ramsay. Two years later the firm was Lennox, Benton and Reynolds. The new partner, Benton, was described as 'Managing Director' in 1903, again suggesting that, as before, the technical skills of the company resided solely in Lennox.

By 1886 Dewar had devised a much better engineered apparatus, for which he abandoned his usual practice of concealing its construction; it was revealed in an abstract of his paper in a newly-founded journal, *Industries*, where it is titled 'Recent researches on meteorites'. This has a detailed engraving and description of the apparatus. Unfortunately the Friday Evening Discourse had the same title, which was not one that was designed to attract the attention of cryogenic enthusiasts.[31] Since meteorites come from outer space, and so from a cryogenic environment, he and Ansdell studied the effects of cooling with liquid oxygen the gases they found in them. The liquefaction apparatus had two large metal cylinders through which passed a copper tube 45 ft (14 m) long that carried the oxygen under a pressure of about 75 atm. In the upper cylinder the tube was cooled by a liquid carbon dioxide/ether solution and in the second or lower cylinder the tube was further cooled by ethylene evaporating under pumping. Liquid oxygen dripped from a valve at the end of the tube. In this way he was able to make up to 15–20

cm^3 of liquid at a time. By further cooling the liquid oxygen by its own evaporation and that of some of the liquid ethylene he could also solidify it and show the solid on the lecture-room screen. The public announcement of the solidification was made on 27 May 1886 by the President, George Stokes, at a meeting of the Royal Society.[31] By 1892 the scale of his apparatus had advanced to the point where he was handling the liquid by the pint (600 cm^3). Meanwhile Olszewski was claiming 100 cm^3 in 1890 and 200 cm^3 the next year. It is probable that the title of this paper on meteorites, and of the Friday Evening Discourse based on it, caused it to be overlooked by Olszewski, in particular, who was later led into a dispute with Dewar over priority because of his ignorance of its contents.[31]

The 20 years from 1878 to 1898 were the most productive of Dewar's career. His liquefaction of hydrogen in 1898, the last of the traditional 'permanent' gases, would have brought his cryogenic work to a triumphant finale had not terrestrial helium been discovered three years earlier by his enemy, William Ramsay. Nevertheless the liquefaction of hydrogen was a splendid achievement, although one that was soon repeated when William Hampson (see below) designed a hydrogen liquefier for Brin's oxygen works in 1900. Ramsay's colleague, Morris Travers, built Hampson's liquefier at University College.[32] Because of Dewar's secrecy about his designs it was Travers's apparatus that was placed eventually in the Science Museum. A larger and revised version of Dewar's apparatus was used at the St Louis International Exhibition in 1904. In Leiden Kamerlingh Onnes (see below) had liquefied air in 1892 and followed with hydrogen in 1906.

Cryogenics was certainly Dewar's principal concern during these two decades but he had other interests as well. The most substantial of these was his spectroscopic work with George Liveing which started in the opening year of 1878 and, as we have seen, led to a flood of publications in the next few years: 41 papers from 1878 to 1883, when the flood abated somewhat. A minor interest from his earliest days was his work on pyridine derivatives which continued in these years. In 1892 he started a colaboration with Ambrose Fleming[33] at University College, London on the electric and magnetic properties of materials, with most emphasis their behaviour at low temperatures. It seems that much of the practical work in these collaborations was done by his colleagues, Liveing and Fleming, and that Dewar and his two laboratory assistants concentrated mainly on the work of gas liquefaction at the RI. He never worked with the large team of students and post-doctoral fellows that modern chemists find necessary for a thriving research programme. This work in fields other than liquefaction will be discussed later.

The theoretical ideas that underlay these experimental advances were few, and not widely understood. In the years 1873 to 1878 Willard Gibbs had re-laid the foundations of classical thermodynamics in a massive paper published in two parts in the *Transactions of the Connecticut Academy of Arts and Sciences* (see Chapter 5) but this work had little influence at the time. This oversight was not primarily because of the obscurity of the journal in which it was published, since he circulated reprints widely throughout Europe, but because of the condensed, formal and abstract language in which it was written. Few physicists and even fewer chemists mastered his work for many years. A second and very different theoretical idea was in the doctoral thesis of Johannes Diderik van der Waals[34] of 1873, but this also initially attracted little attention. Few were inclined to read a thesis in Dutch by an unknown schoolmaster from The Hague; only Maxwell rose to the challenge, and it was his review in *Nature* that ensured some publicity for its remarkable results. Its influence became more widespread only after Friedrich Roth published a German translation in 1881.[34] Van der Waals asked how the existance of forces between molecules changed the equation of state of a perfect gas, $pV = RT$. Here the constant R is proportional to the size of the sample of gas, conveniently taken to be one mole, when we have R is 8.3143 J K^{-1} mol^{-1}. He realised that the pressure is modified by the attractive forces between the molecules and so deduced that p should be replaced by $(p + a/V^2)$, where the parameter a is a measure of the strength of these forces. The molecules also occupy space and so the free volume open to each is less than V; he replaced it by $(V - b)$. The resulting equation,

$$(p + a/V^2)(V - b) = RT, \tag{8.1}$$

is a cubic in the volume and so has three roots at each pair of values of p and T. At high temperatures two of these are imaginary and one is real; it is the real solution of the equation that gives the relation between pressure, volume and temperature that we call the equation of state of the gas. Below a certain critical temperature, $RT_c = 8a/27b$, the cubic equation has three real roots. The largest and smallest are closely related to the molar volumes of the gas and of the liquid with which it is in equilibrium. (The middle root is unphysical and is ignored.) It was by finding the values of p, V and T that marked the transition from one to three real roots that van der Waals located the critical point of his equation of state. There is a simpler way since at this point the first and second derivatives of pressure with respect to volume vanish. The equations $(\partial p/\partial V)_T = 0$, and $(\partial^2 p/\partial V^2)_T = 0$ can be solved simultaneously to give $RT_c = 8a/27b$, $V_c = 3b$, and $p_c = a/27b^2$. This second way was found by Hamilton Dickson in 1880 and is now to be found in every textbook.[35] Van der Waals's equation gives a convincing qualitative account of the behaviour of compressed gases and liquids but was soon found to be inaccurate in its numerical predictions. In 1880, however, he realised that the results above for the expression of the critical constants in terms of the parameters a and b allows these parameters to be replaced by the critical constants themselves. Thus if we write $\pi = p/p_c$, $\omega = V/V_c$ and $\tau = T/T_c$ then the equation becomes:

$$(\pi + 3\omega^{-2})(3\omega - 1) = 8\tau. \tag{8.2}$$

This equation is, of course, no more exact than the one from which it was derived, (8.1), but in 1880 it suggested to van der Waals that a general relation might exist between these 'reduced' variables, as he called them, which was not tied to the specific form of his basic equation. That is, he asked if there might be a general relation between different substances such that for each,

$$\pi = f(\omega,\tau), \tag{8.3}$$

but without any commitment that the specific form of this function is that of (8.2); all that was required is that $f(\omega,\tau)$ be the same function for different substances. He found that this was so, at least to a much better accuracy than any group of substances conformed to (8.2). Thus the critical ratios, $(pV/RT)_c$, for the simple gases argon, nitrogen, oxygen and methane are, respectively, 0.28, 0.29, 0.30 and 0.29, while carbon dioxide has the value 0.31. These numbers are much closer to each other than any is to the value of 3/8, or 0.375, predicted by the van der Waals equation, (8.1) or (8.2). This new result he called the 'law of corresponding states'.

Before he could get this result into print, a young physicist with whom he had discussed it, Heike Kamerlingh Onnes, had deduced that the existence of such a relation implied a corresponding similarity, for different gases, in the form of the function relating the intermolecular forces to the separation of the molecules.[36, 37] Kamerlingh Onnes was then an assistant at Delft Polytechnic to Johannes Bosscha, van der Waals's former mentor, but in 1882 he became Professor of Experimental Physics at Leiden, where Lorentz was the Professor of Theoretical Physics. Here Kamerlingh Onnes remained until his retirement at 70 in 1923. In some ways he resembled Dewar, for both stressed that physics must be founded on quantitative experiments. In the cryogenic laboratory that he established at Leiden he had the phrase *Door meten tot weten* ('Through measurement to knowledge') inscribed at the entrance. He had, however, a serious interest in the theory of what he was doing which contrasted with Dewar's casual attitude to theory, calling on Hamilton Dickson only when he found that he needed advice. Kamerlingh Onnes was also more careful in establishing a structured approach to the training of his technical assistants, a matter that Dewar left to chance, being fortunate in his early choice of Robert Lennox. Neither took kindly to teaching students, both insisted on strict observance of the laboratory rules, both sufered from poor health, and both had terrible handwriting. The two men never met; Dewar did not go to Leiden and Kamerlingh Onnes's only visit to London was in 1924, shortly after Dewar's death, and not long before his own. Most of their correspondence is on the liquefaction of hydrogen and helium but they had an earlier contact over a matter of safety, when, in 1896, Kamerlingh Onnes's laboratory was threatened with closure. The city of Leiden had a horror of explosions that stemmed from a disastrous occasion in 1807 when a French ship laden with gunpowder caught fire and exploded. When the burghers heard that Kamerlingh Onnes had compressed gases in cylinders (including hydrogen) they again feared for their safety. They were supported by the Professor of Anatomy whose laboratory adjoined the Physics Laboratory. The complaint reached the Minister of Internal Affairs who asked the Academy of Sciences for its advice. A committee under van der Waals was set up and sought evidence of how the problem was dealt with elsewhere. The year before, the Royal

Society in London had asked a committee, on which Dewar had sat, to look into the safety of cylinders of compressed gases.[38] The Amsterdam committee asked the opinions of Olszewski in Krakow, Pictet, now in Berlin, and Dewar in London. Kamerlingh Onnes clearly knew of Dewar's secrecy over the construction of his apparatus, for Dewar had written to him on 20 July 1895:

> I trust we may meet some time here, but in the present condition of low temperature discussion I have felt it reasonable and proper to decline allowing even Pictet to inspect my laboratory. The object is to prevent further [recriminations?] seeing the details of my apparatus have not been published.[39]

Kamerlingh Onnes replied:

> Now you wish me this summer that if I come to London you would decline me to inspect your laboratory. Though I did not intend at that time and though I will not have the opportunity before long to come to London, I can say that this communication is a great disappointment to me. I do not think you go as far in secrecy that you will not assist a fellow-worker who is assailed in such a manner as I just communicated to you. So I beg to have to ask you the following questions … .[39]

There were nine questions to which Dewar returned brief but helpful answers saying that there had never been any objections to his operations, that the cylinders of compressed gases were kept below and in the lecture room, that his gas-engine was of 100 horse-power [75 kw], that his ethylene cylinders contained between 60 and 70 lb [25 to 30 kg], that there were 20–30 'bottles' of other gases, that he used no nitrous oxide,[40] that the pressures in his cylinders were from 150 to 300 atm, and that no special precautions were taken 'beyond the ordinary care of scientific men'. Dewar's accompanying letter was particularly forthright:

> It would be a terrible disaster for science in your country (and universal science) if the Municipality of Leiden succeeded in carrying out any restrictions on your splendid cryogenic laboratory and the fine work you are doing. I cannot understand such a position … I may say that I have made all my experiments with high pressure apparatus before the Prince of Wales and the Sister of your Queen Dowager the Duchess of Albany without the slightest hesitation and no suggestions of danger were even suggested.[41]

Fortunately van der Waals himself had come to London five years earlier to hear Rayleigh lecture and to see Dewar's apparatus for himself. This was on the occasion of the celebrations of the centenary of Faraday's birth when the RI decided to bring up to date the depleted ranks of its foreign Honorary Members with 20 new elections. In a carefully apportioned list, five Frenchmen, four Germans, three Americans, two Italians, and one each from Denmark, Sweden,

Russia, Switzerland, Belgium and the Netherlands were invited; the last was van der Waals. The balance of the list reflects the contemporary assessment at the RI of the strengths of the different countries in the (mainly) physical sciences. With Dewar's letter and van der Waals's own evidence, the Academy committee recommended in May 1896 that Kamerlingh Onnes be allowed to continue. It was, however, probably as well that they had reported before they knew of the safety record of the RI laboratories in that year.[41]

8.2 The Vacuum Flask

We must now consider the problem that the work of the pioneers, including Wróblewski and Olszewski, had left unsolved; how to keep for any length of time the liquid oxygen, nitrogen and air that they and he had produced with such difficulty, and at such expense. The solution of this problem proved to be Dewar's most original contribution to the field, the invention of the silvered vacuum flask that now carries his name in laboratories throughout the world. He was rightly proud of this invention which he regarded as the core of his low temperature work. When Kamerlingh Onnes wrote to him to say 'I am sure your vacuum glass is the greatest advance in low temperature work since 1883' Dewar underlined the sentence.[42]

Heat can be transferred to the contents of a closed vessel by three different mechanisms: convection, conduction and radiation. The first is the flow of the atmospheric or other gas around the vessel which brings a continuously changing stream of warm gas to the outer wall of the vessel. This is the easiest problem to solve since it is possible to reduce the pressure of the surroundings to a point where the convective gain of heat is negligible. Even a stationary gas will, however, transfer heat by conduction and this is a more difficult problem. At first sight it might be thought that the thermal conductivity of a gas would fall steadily as the pressure is reduced, but this is not so. As the pressure falls there are fewer molecules to transmit the energy (or heat) but there are also fewer molecules to hinder the molecular motions and so any 'hot' molecule travels further before giving up its energy on collision with a colder one. At and around atmospheric pressure the two effects compensate each other and the conductivity is almost independent of the pressure. (There is a similar effect for the transfer of momentum in a gas and so its viscosity is also independent of pressure at and around atmospheric pressure. It was Maxwell's surprising discovery that this is so which provided one of the first and most powerful arguments in favour of the kinetic theory of gases.) Clearly this compensation cannot persist down to zero pressure and the thermal conductivity and viscosity fall when the mean-free-path of a molecule is comparable with the width of the gap over which the transfer of energy or momentum is being studied. For nitrogen and oxygen, and so for air, the mean-free-paths at atmospheric pressure and room temperature are about 6.10^{-8} m, and so a gap of, say, 1 mm is comparable with the mean-free-path at a pressure of 6.10^{-5} atm, or 0.1 mmHg. To reduce substantially the transfer of heat by conduction requires, therefore, a vacuum pump that can achieve pressures better than 0.1 mmHg. Transfer of heat by radiation is almost independent of the pressure of the surrounding gas and determined by the nature of the surfaces of the bounding solids. 'Silvered' or other highly polished and reflecting surfaces greatly reduce the transfer of heat by radiation.

Most inventions have a disputed history and the vacuum flask is no exception. George Gabb,[43] a chemist and collector of scientific instruments, championed a 'firebox' proposed by Lewis Gompertz, who, he said, was the first, in 1850, to propose that a metal vessel with polished walls, surrounded by a vacuum, would be well-insulated. Gompertz did not, and could not have made such a vessel since the vacuum pumps available in 1850 were inadequate to reduce the pressure sufficiently to lower the effective conductivity of the gas surrounding the vessel to be insulated. The best that a piston air-pump could then achieve was about 3–4 mmHg. A better vacuum could be achieved in the space above a column of mercury of 76 cm or more, the so-called Torricellian vacuum, but it was almost impossible to use such a space for any practical purpose. In 1865, however, Hermann Sprengel, a German working in London, designed a continuously operating pump that worked on essentially Torricellian principles.[44] He poured mercury in drops down a tall vertical glass tube. Between each pair of drops some gas was trapped and so carried to the bottom of the glass tube from where it could be emitted to the atmosphere. Such a device can, in principle, reduce the pressure to that of mercury vapour at room temperature, or about 10^{-3} mmHg, although in practice this limit is never reached. Nevertheless it was Sprengel's pump that first allowed the heat transferred by conduction to be reduced substantially. Dewar used a modified form of his pump in 1872–3 when he wanted to reduce the transfer of heat to a calorimeter in which he was measuring the heat capacity of hydrogen absorbed in palladium.[44, 45] Similar experiments were made in 1881 in Germany by Adolf Weinhold, who was describing a device to exhibit the solidification of mercury, and in 1882 in France by Jules Violle for calorimetry.[46] Weinhold's invention gave his name to the vacuum flask in Germany, when the rest of the world chose Dewar's. In France it was sometimes claimed that Cailletet was the first to use the vacuum flask for keeping liquid air. He himself never seems to have advanced his case although he was present when the claim was made in 1895 by the Belgian chemist Ernest Solvay who was attending a meeting of the French Academy.[47] Cailletet apparently made no move to deny it, to Dewar's subsequent annoyance.[48] In 1897 the even wider claim by William Hampson that he used a 'glass vessel protected by a vacuum of the kind invented by Professor Crookes, first applied to refrigeration work by Professor Cailletet of Paris, and improved and popularised by Professor Dewar' apparently passed without comment from Dewar himself.

These early experiments by others had little if any influence on Dewar's invention of the vacuum flask that he exhibited first at the RI on Christmas Day in 1892, and then described publicly in a Friday Evening Discourse the next month.[49] Who first made and silvered these vessels is a question that is difficult to answer. Dewar was an excellent manipulator of delicate equipment but he was no glass-blower and so the flasks, although designed by him, were made by assistants or by outside workers under contract. The candidates include:

(1) Robert Lennox

William Bragg's account at the RI of the 'History of the Vacuum Flask' was published in *Nature* in 1940, and attracted comment from Lord Rayleigh.[50] This, in turn, caused William Green, who was a research assistant and later lecture-room assistant at the RI from 1900 until 1950, to put the case for Lennox. He wrote to Rayleigh:

> As Dewar's assistant in succession to Lennox and as junior and friend of Lennox in the RI laboratories, I was, of course interested in your letter in *Nature* concerning the origins of the vacuum flask. Sir Wm Bragg [the Director of the RI when Green was writing] tells me I am free, as far as he is concerned, to say what I feel in the matter, which is of deep interest to all of us at the RI. Mr Heath told me quite definitely that Lennox made the first vacuum vessels, and further he passed on his methods to outside people who were asked to supply more than the laboratory could make. I am sure that Sir Wm has this view also from the discussions with me during the preparation for the December lectures; indeed we had one of Lennox's own vessels ('the first Hg.-mirror vac. vessel') displayed with the collection of typical vacuum vessels at the discourse; this vessel at least I am certain of, as having been identified by Lennox directly to me: so nobody here has any doubt of Lennox's position in the matter: but Dewar certainly called in outside aid for a supply of vessels, and Mr. Gimingham, of Edison & Swan, was certainly one of these persons. I don't know whether such people knew of Lennox's work; certainly in the letter written by Dewar to Gimingham no reference is made: but as you are probably aware, Dewar was well able gather aid & information from many sources without always bringing them into touch with each other; and it is just possible that some of these people came to consider themselves as prime movers.[51]

Dewar had commended Lennox for his skill in glass-blowing at the end of a paper of 1888.[52]

(2) Edward Gimingham

Edward Alfred Gimingham was the brother of Charles Henry Gimingham, the research assistant of William Crookes. Charles moved in 1885 to the newly amalgamated Edison and Swan Electric Light Company (or Ediswan as it became known) and asked his younger brother to join him there; Edward was then only 18 or 19. Charles was an expert glass-blower who had made the radiometers for Crookes, and it was presumably he who instructed Edward. Charles died in 1890 at the age 37, but Edward lived on to become a Member of the Royal Institution in

1896, a director of the Ediswan company, and died only in 1953.[53] Dewar wrote to Edward just before the display of the first vacuum flask at Christmas 1892 to say 'your beautiful bulb arrived … if you come on Friday evening on Jany 20th you will hear your name praised. … Then I would like one or two coated on the outside with Silver!'.[54] But did Dewar actually mention Gimingham at the Discourse? His name is in a report on the lecture that appeared in *The Engineer*: 'Mr E. Gimingham had, at great inconvenience and with much skill done the glass-blowing for him.' We do not know how far this report was based on what their representative had heard at the lecture and how far on a summary that may have been supplied by Dewar himself.[55] The more complete account that Dewar later published makes no mention of Gimingham.[56] Perhaps by this time Lennox was making the flasks and Dewar might not wish to diminish his credit in the enterprise.

It was presumably Edward Gimingham that Ambrose Fleming had in mind when he wrote to Armstrong on 14 May 1923, after Dewar's death, with material for the obituary that Armstrong was writing. Fleming wrote:

> The first paper [on the resistance of carbon filaments at low temperatures] published by us in 1892 was preliminary and the difficulties of keeping constant the temperature turned his [Dewar's] thoughts to the vacuum vessel & he gave me a rough sketch one day of such a double walled tube exhausted and asked me to have one made at the Edison and Swan works, which I did. Afterwards his assistant Lennox began to make and silver them. That was the first vacuum tube.[57]

(3) Charles Muller

George Gabb put forward the claim of Muller, a German-born glass-blower in private practice, in his correspondence with William Bragg at the time of Bragg's two lectures at the RI in December 1939, and in Gabb's note in *Nature*.[50] He had previously written to the *Daily Telegraph* (12 October 1935) to advocate the claims of Gompertz, but now in a second letter, which the paper declined to publish, he wrote:

> ~~The statement that the first Vacuum flask was made by the Edison and Swan Coy at their factory is contrary to my personal knowledge.~~ To my personal knowledge, the first double-walled glass Vacuum flasks were made on Dewar's instructions by that expert glass-blower Charles Muller of Holborn, with whom I was collaborating in certain chemical glass apparatus at the time. ~~For several months in 1892 all the~~ The first experimental Vacuum ~~vessels~~ flasks he made for Dewar in 1892 were all made of clear glass. ~~It was~~ after a considerable number of these had been made Dewar then suggested that the internal surfaces should be silvered, and as he [Muller] had no experience of this I instructed him in the technique of silvering several of the ~~first three or four~~ flasks for him, and these were undoubtedly the first Vacuum flasks ever silvered. Specimens of these are

> now preserved in the Museum of the Royal Institution. The above statement can be confirmed by Mr. C.L. Muller, a cousin of the late Charles Muller, who was associated with him at the time, and who still carries on the business of a Scientific glass blower in Parton Street, near Holborn.[58]

The conflict of evidence in these testimonies is, perhaps, best resolved by discarding Green's second-hand claim that Lennox was the first to make the vacuum flask, but to accept his remark that Dewar often ran several horses at the same time. Gabb's evidence is at first-hand and he seems to have no motive in inventing anything, while Dewar's letter to Gimingham is equally explicit and is supported by Fleming's memory. It seems most likely that Dewar, perhaps not trusting the ability of either Muller or Gimingham to succeed at once, approached both in 1892 and asked them to make the flasks. The maker must have been paid, but there is no record at the RI of any payment to Muller or to Gimingham; Dewar could, however, have met any bill of Muller's from his own pocket. A payment to Gimingham would presumably have gone through his employer, Edison and Swan, and here there is a suggestive line of payments. The company was formed in 1883, Edward Gimingham joined it in 1885 or 1886 and the first payment to the company was in 1887, followed by another in 1888. There was only one payment in the next five years but then a steady flow of payments from 1893 onwards, with peaks in 1896 and 1898 when Dewar was involved first in developing his apparatus for gas liquefaction by Joule-Thomson cooling in 1895, and then its exploitation for the liquefaction of hydrogen in 1898. In 1901 and 1902 there were again no payments. Vacuum flasks may not have been the only equipment bought from Edison and Swan, indeed one small payment of £4 is specifically said to be for switches, as if, perhaps, to mark its singularity, but it is probable that most of these payments were for flasks made by Gimingham, or, later, perhaps by one of his colleagues at the company.[59] The verdict that Muller and Gimingham were both involved in the first experiments was reached also by William Bragg.[50]

Dewar measured the efficiency of his flasks by noting the rate loss by evaporation of liquid air. No doubt flasks of different shapes behaved differently and the figures that he quotes vary over the years. By the time of his Friday Evening Discourse on 27 March 1896 his results show that if he started with a double-walled vessel then evacuation of the space between the walls cuts the rate to about 20 per cent of its previous value. Silvering both internal walls reduces the rate further to about 4 per cent.[49] (As usual he received theoretical advice from his brother-in-law, in this case on the physics of heat transfer.[60]) As an alternative to chemical silvering, as practised by Gabb and Muller, Dewar showed that it was equally effective to 'silver' the inner wall of the evacuated space by having a small drop of mercury at the bottom of the space. This condensed on the cold wall when the liquid air was placed in the flask. It was comparatively straightforward to make a cylindrical vessel from two glass tubes of appropriate diameters, but it was much harder to make a spherical flask since a tube of very large diameter had to be heated and bent *in situ* into a spherical shape so that it could be sealed to the inner sphere at the neck and then drawn down to a small diameter at the base to permit the evacuation. Lennox mastered this manipulation with the help of three other men who held the semi-molten glass in the right position and shape. All this work was carried out with 'soft' or soda glass, before the days when the harder borosilicate glass, with its lower coefficient of thermal expansion, became available. Each flask had therefore to be carefully annealed to release the stresses of the manipulation before it could be left to the final cooling by itself. Quartz has a lower coefficient of expansion than glass but was found to be less efficient than soft glass as a material for the walls of the vessels.[61] The critical area in all flasks was the large joint at the top of the vessel which was prone to shatter when the liquid air was poured out. The danger was lessened if the seal was made not at the lip but lower down on the surface of the outer cylinder. One particularly complex flask was designed to deal with this problem by the insertion of a small helical internal tube that led from the bottom of the flask to the outside surroundings through which the liquid air could be extracted; such flasks were made for Dewar in Berlin and may have been first used in the apparatus for the liquefaction of hydrogen in 1898 (see Figure 8.3).

Dewar apparently used his flasks only to keep liquids cold; he presumably thought that this was a specialised use for which the market for the flasks would be small and he never patented the apparatus or its mode of manufacture. Had he done so he would no doubt have become an even wealthier man. It was a German, Reinhold Burger, who first thought of using a vacuum flask to keep his coffee hot, or his baby's milk for a night feed – the accounts vary. He took out a German patent, set up a company, and sponsored a competition for the best name for his 'invention'. This was won in 1904 by a Munich resident who proposed the name 'Thermos flask' which was then registered as a trade mark. In 1907 a group of business men acquired the British rights and set up the company 'Thermos Ltd.'. When, in 1904, Dewar made the flasks out of metal and had found that charcoal cooled in liquid air absorbed any gas retained on the surface of the metal then he did take out a patent but because of the complications over his rights in Germany he never reaped much reward from it. So in 1918 the court allowed him an extension of a further four years, but it is unlikely that he profited much from this during his last few years.[62] Meanwhile the flask had become known as a Dewar bulb or flask throughout the scientific world. The former words were used in America by Thomas Sloane in 1899, within six years of its invention, in his book on low temperatures,[18] but they must have become established before then since an advertisement at the back of his book by the Edison Decorative and Miniature Lamps Department of the General Electric Co. claims that 'We were the first in this country [USA] to make Dewar bulbs for preserving liquid air'. An advertiser is not going to offer for sale an article whose name is not already well-known to its customers. The flasks were not available commercially in Britain until the twentieth century since William Ramsay had to send to Berlin for his flasks. The noun 'dewar' followed later but has not been dated precisely. The non-scientific world, however, soon adopted the name 'Thermos', and Dewar himself could say in a Discourse in 1914: 'Such vessels are commonly called Thermos flasks'.[63] In spite of his apparently untroubled acceptance of the adoption of his invention in this way, he cannot have been best pleased to receive an enquiry about a patent from the British company Thermos Ltd. in January 1909 asking whether, in Germany in 1903, the phrase 'Dewar vessel' or 'Dewar-Weinhold vessel' referred solely to vessels made of glass or whether they included metal vessels also. Dewar had little liking for Germans and probably did not read the language. He replied rather testily, and surely not quite truthfully, to say that he knew neither phrase, nor what they meant in German.[64]

The extension of vacuum vessels from glass to metal was an important matter, for more and more of high pressure and low temperature apparatus was being made of metal which was stronger and more easily worked than glass. The problem was that on evacuation a metal vessel did not 'hold' its vacuum which was steadily destroyed by the slow release of gas adsorbed on the surface of the metal. In a preliminary note to their paper on the working of Crookes's radiometer Tait and Dewar had found a way of improving the vacuum obtained by the best pumps. Its title, 'Preliminary note on a new method of obtaining very perfect vacua', recalls that of an earlier paper by Thomas Andrews,[65] in which he used the trick of replacing the air in the vessel to be evacuated by carbon dioxide, absorbing the last traces of that by potassium hydroxide, and then absorbing the water vapour with concentrated sulphuric acid. Tait was collaborating with Andrews at the time of that work and must have had Andrews's work in mind when he and Dewar devised a simpler way of improving a vacuum. They enclosed some coconut charcoal in the vessel which they pumped down, while heating the charcoal to as high a temperature as the glass would stand. They then sealed the vessel and let the charcoal cool, when it absorbed the remaining gases. Their criterion for the quality of the vacuum was the ease with which it passed an electric spark between two metal electrodes sealed into the walls of the vessel. They found that no spark passed at atmospheric pressure but at a pressure of a few mmHg the spark passed easily. On improving the vacuum further no spark could be passed. Such insulation was a simple test for the quality of a vacuum that was used for many years in laboratories round the world. The insulation was improved when liquid air, and later liquid hydrogen, became freely available since charcoal cooled with these took up, and retained, all gases.[66] Its use solved the problem of getting and keeping a good vacuum in a metal vessel since the cooled charcoal continued absorbing the gases given off by the metal. J.J. Thomson claimed that such vessels made a key contribution to his work on the properties of the sub-atomic particles. Dewar became most interested in the properties of the charcoal and of the phenomena that it could be used to study; it formed the focus of many of his researches in the years after 1904. This work was introduced in a Friday Evening Discourse which opened with a historical account of work on the occlusion of gases by solids.[67]

8.3 Interchangers

Dewar always maintained that his principal aim was the study of the properties of matter at low temperatures, and that the attainment of such temperatures by the liquefaction of the 'permanent' gases was only a means to this end. There were therefore many experiments from the 1880s onwards in which liquefaction played no essential role. He recognised, however, that as long as hydrogen remained unliquefied there were temperatures that were too low for him to explore, and his thoughts could never have been far away from seeking a solution to this 'last' problem. The critical temperature of oxygen is at 155 K and nitrogen a little lower at 126 K, but that of hydrogen could not be measured until its liquefaction had been achieved. It was, however, estimated by Wróblewski by measuring the isotherms of the compressed gas, that is, the change of pressure with volume along a line of constant temperature, and then comparing these with those of nitrogen or oxygen by means of the law of corresponding states of van der Waals. This comparison led him to suggest that the critical temperature of hydrogen was below 40 K. The gap between 40 and 126 K was too wide to be bridged by any other substance that might have served as a stepping-stone in a cascade process of the kind that had been so successful in getting from room temperature down to 100 K, or thereabouts, by using carbon dioxide, nitrous oxide and ethylene. Dewar was therefore led to wonder if a mixture of nitrogen and hydrogen might serve, since Ansdell had shown that the critical temperature of a mixture is roughly linear in its composition. But this suggestion did not work. He prepared a paste of solid nitrogen and gaseous hydrogen with which he claimed to have reached a new record for the lowest temperature in a laboratory, but he could not make liquid hydrogen in bulk.[68] Some radically new ideas were needed if this gap in the cascade process were to be closed. These were eventually forthcoming from improvements in technique arising from the refrigeration industry and from the exploitation of an effect discovered many years earlier by two scientists who had no practical applications in mind, the Joule-Thomson effect.

The shipping of meat from Australia started in 1879 when the steamer *Strathleven* brought frozen meat to Britain. The cooling of the cargo was based on what were by then well-established thermodynamic principles: it required (1) cooling a compressed gas from ambient temperature by making it do work by expansion; (2) using the expanded gas to cool the meat, so warming the gas back to ambient temperature; (3) recompressing it so that it became hot; and then (4) cooling it back to ambient temperature again with cold water, so that it could go round the cycle again. The net result of this refrigeration was the transfer of heat from the meat in the second stage of the cycle, to the cold water in the fourth. The cost of this process depends on the efficiency of the transfer of heat at each stage and so improving this efficiency attracted the attention of the engineers in a more serious way than had the small-scale operations of the academic scientists seeking to obtain a few litres or so of a liquefied gas. The most important device in transferring the heat was what was called a regenerator or an interchanger;

the names were often used indifferently, since they achieve the same end, but it is convenient to distinguish them here. Both were devices that aimed to transfer heat from one stream of gas to another in a reversible way, that is, in a way that is completely efficient from a thermodynamic point of view.

The regenerator was introduced by the Rev. Robert Stirling in his heat engine of 1816, but his work was ignored for many years and it was William Siemens who deserves much of the credit for the widespread use of regenerators from the middle of the century onwards. In its simplest form it is a thermally insulated tube packed with a material of high thermal conductivity and large surface area, such as wire gauze. On passing a hot gas into the tube from, say, the left-hand end, the gauze will take up heat from the gas until there is a gradient of temperature along the tube from that of the hot gas at the left end to the ambient temperature at the right-hand end. The gas stream is now stopped and a second stream of gas passed down the tube in the reverse direction. It acquires heat from the gauze until it emerges at the left end with, in an ideal case, the temperature of the original hot gas as it had entered the tube. Throughout the operation the transfer of heat takes place between gauze and gas only at places where they are in contact at temperatures that differ by infinitesimal amounts, so that the transfer is almost reversible in a thermodynamic sense. Siemens first discussed the operation of such a regenerator in a report and provisional patent application of 1857 and his principal use of it, from which he made much of his fortune, was from 1861 in the operation of open-hearth steel-making plant.[69] In a simple furnace cold gas is used to burn a fuel and the products of the combustion are rejected to the atmosphere as a hot gas. Siemens improved the efficiency of this process by interposing a regenerator acting between the incoming cold gas and the outgoing hot gas of combustion. The result is that the gas that is about to be burnt is already hot. After being burnt the gaseous products are returned to a second regenerator where they transfer their heat to an incoming stream and so leave the plant as a cold gas. The process is now more efficient since the gas used in the combustion is already hot, and no heat is wasted in the discharge of the now cold products of the combustion. Such regenerators were applied for cooling rather than heating by Alexander Kirk[70] in 1862 and later by others also.

For the liquefaction of gases it proved simpler and better to use what is usually called an interchanger, although Dewar himself always retained the name regenerator for it. This comprises two long tubes in good thermal contact throughout their lengths. This contact can be established in many ways, the simplest perhaps being to thread one tube inside the other. If now a stream of hot gas is passed into, say, the left-hand end of the inner tube and a cold stream of gas into the right-hand end of the outer tube then, when a steady temperature gradient has been established, and if the tubes are long enough, it is found that hot gas comes out of the left-hand end of the outer tube and cold gas from the right-hand end of the inner tube. The thermal contact between the two streams is, at every point along the two tubes, between gases at almost the same temperature, so that the transfer of heat from the inner stream to the outer is again carried out reversibly with, in

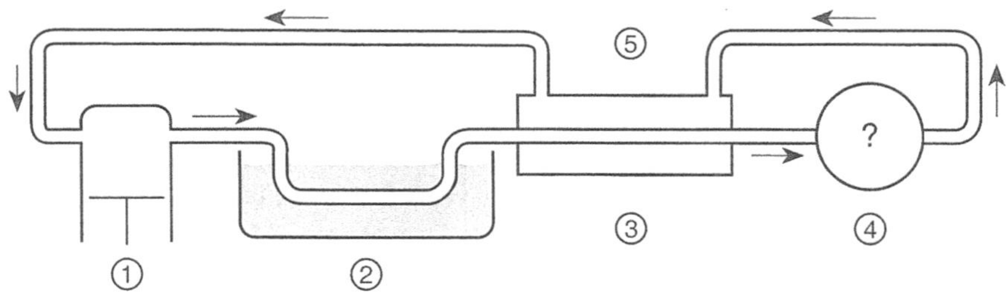

Figure 8.1 Schematic sketch of the cycle in an apparatus for the cooling of a gas.

principle, complete efficiency. Siemens had proposed the use of an interchanger as an alternative to the regenerator but he never put it into practice for refrigeration.

The essential step in reaching low temperatures in a usable way is the extraction of energy from the gas. This can be done in a process in which a stream of gas is first compressed and so heated, then cooled back to its original temperature by passing it through a bath of cold water, then expanding the gas by making it do work against a piston in a cylinder so cooling it further, and finally returning it to the original compressor. On each cycle the gas gets cooler and, in principle, the process can continue indefinitely until brought to a halt by some outside problem, such as that of lubricating the pistons. Joseph Coleman used such a scheme in 1879 which he described to the Institution of Civil Engineers in 1882, and again, with John McKendrick, in a Friday Evening Discourse at the RI on 29 May 1885.[71]

The process is made efficient by the insertion of an interchanger, as is shown schematically in Figure 8.1. The analysis of the operation can now be described, as follows. At stage (1) the gas is compressed and so heated. This heat is removed by cooling with cold water in stage (2). The gas enters the inner tube of the interchanger and is cooled further by the stream flowing the other way in the outer tube (3). At stage (4) the pressure is reduced by expanding the gas, an operation which can occur in different ways. If the gas is made to do work by driving a piston in a cylinder or by turning a turbine then it is cooled further and passes back into the interchanger at stage (5). It is, however, now colder than the gas flowing into the interchanger and so it emerges at the end of stage (5) colder than the original gas which started at stage (1). There is thus an imbalance in the whole process with the energy put in as mechanical energy in (1) ending in the cold water in (2), and some extra energy being removed at stage (4). This extra energy can come only from the gas, which is thereby cooled. As before, the process continues indefinitely with the gas becoming ever colder in each cycle until further cooling is brought to a halt by some outside problem.

This addition of the interchanger to the scheme used by Coleman and others was proposed by Ernest Solvay in patents in several countries in 1885–7.[72] His system worked well and the temperature went down steadily but eventually came to a halt for the practical reasons that had nothing to do with the thermodynamics of the process – the difficulty of insulating thermally his expansion piston and cylinder, and the difficulty of their lubrication. The lowest temperature that he could reach was about 178 K, which is above the critical temperatures of both oxygen and nitrogen. So the solution of the problem of liquefying a gas without the external cooling of a cascade process depended on finding another way of cooling the gas in stage (4) which overcame the problems of insulation and lubrication.

8.4 Joule-Thomson Cooling

There are three recognised ways of expanding a gas adiabatically, that is, without transfer of heat to or from the gas. The first is to expand it by making it do the maximum work possible, as, in principle, is done in Coleman's and Solvay's processes above. This expansion produces the greatest possible cooling but gives rise to practical problems at low temperatures. The second is an expansion in a static system in which no work is done. Such an expansion is associated today with Joule's name and it played no role in machines for liquefying gases; it is mentioned here since in early discussions such expansion was often confused with the third method of expansion, in which no work is done by a flowing gas, the Joule-Thomson process. The confusion is made worse by the occasional use at that time of the term 'Joule effect' for what we now call the 'Joule-Thomson effect'; it was the latter which was to play an important role in gas liquefaction.

In the 1850s neither Joule nor Thomson was interested in gas liquefaction but they were studying, first, the relation between the absolute scale of temperature and that measured on a gas thermometer, and, second, the foundations of the first law of thermodynamics, namely the exchange of heat and work at a fixed rate of exchange.[73] The credit for the discovery of the first law is usually divided between Julius Mayer and James Joule. Mayer, a German physician, arrived at it by theoretical arguments which depended on the assumption (in modern terms) that the energy of a gas is determined solely by its temperature and is independent of its density. Joule arrived at the first law by well-designed experiments in which mechanical work was turned into heat by the vigorous stirring of a tank of water. He was, however, anxious to see whether what he and William Thomson always called 'Mayer's hypothesis' held true. To this end he took two similar heavy steel cylinders, each of about a litre capacity. He filled one with compressed air, evacuated the other, and connected them by a tube with a valve. This combination was then immersed in a water-bath, the temperature of which he could measure precisely. On opening the valve the gas flowed into the second cylinder and so the gas density was reduced by half. He could detect no change of temperature in the water-bath and deduced that Mayer's hypothesis was confirmed. Later, however, he realised that the great heat capacity of his two cylinders and of the water meant that any modest change of temperature of the gas would go undetected. A similar problem had arisen in early attempts to determine the heat capacity of a gas by heating it in a sealed vessel; the heat capacity of the vessel was so high that the heat capacity of the gas could not be measured with any accuracy. This problem had been overcome by heating a gas flowing down a tube, for when a steady state has been reached the rise in temperature of the gas is independent of the heat capacity of the tube and depends only on that of the gas. It was probably William Thomson who suggested that the testing of Mayer's hypothesis could be better done by expanding a steady stream of gas through a valve and seeing if the temperature changed. The experiment is still a difficult one since it is hard to insulate the tube and valve from the heat of the surroundings, and to maintain a steady flow for a long time of any but the cheapest of gases without elaborate circulating pumps. For 10 years from 1852 Joule worked on this problem and Thomson calculated how the small observed changes of temperature were reflected in departures of the gas from the perfect gas laws.

The results of the three methods of expansion can be compared by considering the cooling of nitrogen from a pressure of 10 bar and at a temperature of 0 °C:[74]

Expansion with performance of maximum work	10.80 K/bar
Joule's 'static' expansion	0.15 K/bar
Joule and Thomson's expansion of a flowing gas	0.20 K/bar

These figures show clearly the great efficiency of the first expansion, as was pointed out by Rayleigh in 1898 who suggested that the practical problems of lubricating a machine capable of extracting the 'maximum work' might be avoided if the machine were a turbine on a long shaft in which the bearings were well-removed from the point of gas expansion.[75] The second and third figures in the list above are zero for a perfect gas, and small and positive for nitrogen at ambient temperatures. We now know that they depend, in a far from simple way, on the weak intermolecular forces between gas molecules. At most temperatures it is the effect of the long-range attractive forces that determine that the expansions lead to cooling, as the average separation of the molecules is increased. At high temperatures, however, the repulsive forces between the molecules become more important and the Joule-Thomson expansion produces heating; that is, the coefficient shown above becomes negative. The temperature at which this change occurs is called the 'inversion temperature'. It is, by van der Waals's law of corresponding states, roughly proportional to the critical temperature. His equation of state requires that the inversion temperature at low pressure is 27/4 of the critical temperature, which would be 850 K for nitrogen. This is too high; the observed value is about 620 K, or about 450 °C. This is still well above ambient temperature and so there is no problem in cooling nitrogen by Joule-Thomson expansion. The gases air (0.25 K/bar) and oxygen (0.40 K/bar) behave similarly. For hydrogen and helium, however, the critical temperature and the inversion temperature at low pressure are below ambient temperatures, namely 33 K and 193 K for hydrogen, and 5 K and 51 K for helium. It is clear that preliminary cooling to below 193 K or 51 K respectively by some other method is needed before the Joule-Thomson effect can be exploited for liquefying these gases. With this introductory account of the ways of cooling a gas by expansion we can now turn to the history of how things turned out.

In April 1894 William Hampson[76] drew up plans for the liquefaction of air. How and why he came to embark on this venture is not known. He was apparently a man of some independent means, a graduate of Trinity College, Oxford, whose degree was in 'Classical Greats', that is, in Latin, Greek, and ancient history and philosophy. After graduation he turned to the law and trained as a barrister, but apparently never qualified. It seems that he had no formal instruction in physics or engineering, and indeed his theoretical knowledge of science was probably minimal. He proposed to use essentially the cycle above with an interchanger in which the same stream of gas passed through the two channels but in which it was cooled by expansion through a valve between the two passages. He made no provision, however, for extracting heat at stage (4) by making the gas do work, as Solvay had done. Those who knew nothing of the Joule-Thomson effect, or who believed it too small to be relevant, would say that without this provision there would be no net cooling of the gas; it would arrive back at stage (1) at the same temperature as it had originally entered the compressor there. This is the behaviour expected from a perfect gas.

Before attempting to put this new idea into practice he went to the RI in November 1894 and discussed it with Robert Lennox. He proposed that he and the group at the RI should collaborate to develop his scheme. Dewar was later to complain of his approach to Lennox rather than to Dewar himself, but Hampson explained that he had had an introduction to Lennox, but did not know Dewar. He came again to the RI and saw Lennox early in December and later said that this call was at about 11 in the morning when he might well have been seen by Dewar in the laboratory, and that he was by no means trying to go behind his back.[77] Nevertheless Dewar, with his obsessive secrecy about the progress of his research, formally reported to the Managers that he had repelled 'endeavours on the part of certain strangers to obtain access to the laboratory'.[78]

Hampson's talks with Lennox got nowhere. At first, according to Hampson's later account, Lennox was attracted to the proposal and asked for time to think about it overnight. The next day he was still interested and suggested that Hampson develop the idea further. Lennox's version of these conversations is that he told Hampson that his idea was 'nebulous' and, on reflection, that it would not work. Hampson said that he wrote three letters to Lennox early in 1895, but when he got no reply he gave up the attempted collaboration[79] and applied in May 1895 for a patent on his proposal. His provisional application consisted of one sentence of 90 words that give no hint that he realised what the problem might be. He wrote:

> The usual cycle of compression, cooling and expansion is modified by using all the gas, after its expansion, to reduce as nearly as possible to its own temperature the compressed gas which is on its way to be expanded; with this object all the expanded gas surrounds the pipe or pipes of compressed gas through all their length from the point of expansion to the point of normal temperature,

> and the length of pipe is sufficient to allow the fullest possible interchange of temperatures between the compressed and expanded gas.[80]

This proposal might have been expressed more elegantly but it seems that the use of an interchanger, which he presumably believed to be a novelty, is intended to be the patentable idea here. There is no hint that there might be a problem with the simple expansion as a source of cooling. The complete application was not submitted until 19 February 1896 but it too contains nothing to suggest that he understood how his proposed apparatus might work. It is unlikely that he had the facilities, or indeed the skill, to build an apparatus himself with which to put his claim into practice. It was to be March 1896 before he was able to demonstrate a working apparatus. He had, by then, established friendly contact with Kenneth Murray of Brin's oxygen company and, although it is nowhere clearly stated, it was presumably Brin's who built the apparatus from Hampson's designs. This company prepared and sold oxygen that was made chemically, usually by forming barium peroxide from barium oxide and air and then decomposing the peroxide to generate oxygen and to recover the oxide.[81] It was an expensive process and so they were attracted by the possibilities of a scheme for extracting oxygen more directly from the air by liquefaction and then the selective removal of the more volatile nitrogen.

Meanwhile Dewar and Lennox had not been idle. In the summer 1894 there is mention in Dewar's laboratory notebook of the 'first large trial of the large apparatus with hydrogen'. This was presumably the apparatus of 1885, as distinct from the small apparatus used previously for his lecture demonstrations. It was tested on 6 August 1894, but no liquid hydrogen was produced.[82] The first mention of a radically new apparatus is in an entry that is not dated but which, from its position in the book, must have been written in about June 1895. It was called 'the direct liquid air apparatus'. On 5 August the 'big apparatus' was used to test how rapidly it could produce liquid air; three-quarters of a litre was obtained in two hours, and in all he 'drew off 5 litres'. A few pages later another undated entry refers to a plan that 'worked well in the Direct Liquefaction of Air Apparatus', which he believes is now sufficiently important to merit capital letters. It was at this time, July 1895, that Dewar wrote to Kamerlingh Onnes to say that he would not allow either him or Pictet to see his laboratory, and, from his comments, to suggest that rumours were starting to circulate about the novelty of Dewar's recent work.

It was unusual for the monthly meetings of the Managers to deal with anything other than the business side of the running of the RI but on 2 December 1895 the new 'direct' apparatus was revealed:

> The Managers after the Meeting went to the Laboratory where Professor Dewar showed the making of liquid air by a simple method. The Professor uses direct expansion of compressed air at ordinary temperature which is made to circulate after expansion from a nozzle over a metallic coil placed between two vacuum vessels.[83]

This demonstration was described in *The Times* the next day:

> A number of men of science witnessed yesterday at the Royal Institution the working of a new apparatus for the production of liquid air with a degree of ease not hitherto attainable. ... Professor Dewar has now produced liquid air by a process so simple and so cheap as to place it at the disposal of every one requiring it for the purposes of demonstration or inquiry. The principle is a familiar one, and has long been practically applied upon a large scale in the laboratory of the Royal Institution. But as expensive gases were dealt with in enormous quantities, the apparatus had to provide for the recovery and recompression after expansion. In other words it had to be of a complicated kind.[83]

The Times reporter, obviously briefed by Dewar, was clearly more impressed by the simplicity and cheapness of the new apparatus than by any new principle involved. Indeed he denies that there is a new principle, an argument that Dewar himself advanced a few days later. The reporter was probably Hugh Munro Ross, who was another classics graduate from Oxford. He had joined *The Times* in 1893 and, like Hampson, he too soon turned to scientific and engineering fields.[84] He became a close friend of Dewar's and assisted in defending him in some of the later disputes in which Dewar became involved. This short report in *The Times* provoked a sarcastic and rather pompous reply from Hampson. He may have sent this originally to that paper, but, if he did, it was turned down and it appeared in the weekly journal *Engineering* on 13 December, to be followed a few days later by a similar letter from him in *The Engineer*.[85]

This narrative raises two questions; did Dewar and Lennox copy Hampson's design for the new apparatus, and did they, and Hampson, understand the new principle involved? Neither question now admits of a totally certain reply but the evidence is strongly in favour of the answers that they *did* copy Hampson, and that *neither* they *nor* Hampson, when they first started on the venture, understood the fact that cooling by the Joule-Thomson effect was essential to the process.

The evidence on the first point is the timing of the visits of Hampson to the RI and of his subsequent letters. It was in November and early December 1894 that he visited the laboratory and in January 1895 that he wrote letters to which he had no answers. Dewar reported to the Managers on 3 December 1894 the attempted visits by strangers to his laboratory, the only time that he ever made such a report. The new 'direct' liquid air apparatus had been constructed by the summer of 1895 at a time when he would not allow Pictet or Kamerlingh Onnes to see his experiments; it was in its finished form and was exhibited privately on 2 December 1895, and was described publicly at a Chemical Society lecture on 19 December. The history of science is replete with examples of the simultaneous and apparently coincidental appearance of the same or similar results from different groups, but the sequence here seems to favour copying rather than coincidence. Moreover, as we shall see, the designs of their two pieces of apparatus, Dewar's and Hampson's, were remarkably similar. Dewar's later behaviour towards Hampson was always to deny that he had learnt anything from him. He went as far as to say that he had never heard of Hampson until his own work was completed.[86] This surely cannot be the case. We know from Dewar's report to the Managers of the RI on 3 December 1894 that Lennox had told him of Hampson's first visits and it seems inconceivable that Lennox did not tell Dewar the name of his visitor. Dewar cited a paper with Hampson's name in a footnote to the later version of his lecture to the Chemical Society that was given as a Friday Discourse at the RI on 27 March 1896.[87] Lennox went so far as to say in 1898 that he had never told Dewar of Hampson's visits until that year. Again it is hard to accept this statement in view of Dewar's report to the Managers in December 1894 and of the extensive discussion, arguments and correspondence on the subject of the liquefaction of gases in the three journals *Nature*, *Engineering* and *The Engineer* in the preceding three years. Dewar, Lennox and Hampson all participated in these discussions. At one point Hampson threatened to sue the firm of Lennox, Reynolds and Fyfe, who had made Dewar's apparatus, for the breach of his patent, but he never followed up his threat, so attracting taunts from Dewar that he had no valid case. Hampson had trained to be a barrister but he wisely refrained from entangling himself in patent law on which Dewar was much more experienced than he was.[88] A comparison of the letters from the three principals in this dispute leaves one with the impression that Hampson's account rings more true than those of Dewar and Lennox and, with the factual evidence of the timing of the developments, certainly leaves me with little doubt that it was only after Lennox had seen Hampson's plans that he and Dewar tried out the method for themselves.

There was, however, a genuine coincidence in that similar ideas were being put forward and similar work was being carried on in Germany at the same time as Hampson and Dewar were contesting the priority in London. The first inkling of the German work to be received in Britain was a brief letter in *Nature* from a Herr Blass at Essen in the Ruhr, published on 29 August 1895.[89] He seemed to have had a common misapprehension, which he may have shared then with Hampson and Dewar, that mere expansion of a gas at a nozzle produces a high degree of cooling, regardless of the nature of the gas or of its subsequent history in its passage through the apparatus. He sketched a device in which a gas was expanded at a nozzle and then circulated back over the incoming stream in an interchanger, although he does not give that part a name. The whole apparatus is thermally insulated from its surroundings. He asks what is the final temperature of the gas emerging from the nozzle if the apparatus is operated for an indefinitely long time. He had drawn a simple form of the apparatus used for cooling by the Joule-Thomson effect although he does not use those words and seems to be ignorant of the effect. He received an answer a week later[89] from Edward Dixon, a young Cambridge graduate, who said that he had heard from his brother, a 'freezing engineer', that such a device had recently been built in Germany by Carl Linde, the proprietor of a well-known manufacturing firm of commercial refrigeration plants.[90] Dixon presumed that the answer to Blass's question was that the lowest temperature that could be reached was that at which the gas liquefied, but then went on to speculate that with a gas that could not be liquefied the limit would be the absolute zero of the thermodynamic scale.

The reference that Dixon gave for Linde's paper was the August 1895 issue of the *Zeitschrift für die gesammte Kälte-Industrie*. This journal of the refrigeration industry was not taken by the RI Library and the British Library holdings do not go back beyond 1910, so it is unlikely that Dewar would have seen this paper. More useful information on Linde's work was fortunately soon available. Herr Schröter described the apparatus before a meeting of German engineers in Aachen, and wrote that Linde had demonstrated his apparatus to scientists and engineers in his works in Munich in May 1895. Schröter makes it clear how Linde's work depended on the cooling by the Joule-Thomson effect. An anonymous translation of Schröter's paper was published by *The Engineer* on 4 October 1895.[91] This Dewar most probably did see since the journal was in the RI Library. Moreover a few weeks later he wrote to Thomson (now Lord Kelvin) from Cambridge, in the rather fawning style he seems to have used in his letters to Kelvin, the most senior of the British physicists:[92]

> Private 29th October 95
>
> Dear Lord Kelvin,
>
> I have just received the results of some experiments which I left going this morning. And I think they will interest you. For long I have been at the Hydrogen problem and have almost given it up in despair, as beyond my resources. The result has been that I have had to invert the problem and ask myself how can air be liquefied from temperatures far above the critical point. This I find is easy if I repeat the old experiment you made with Joule. Thus I find that liquid oxygen can be collected directly from the gas 40° above its critical point and liquid air 67° above its critical. ... Thus I feel almost certain that if hydrogen in the liquid state can be collected I will do it through and on account of the work of Kelvin and Joule.
>
> Ever Yours Truly
>
> James Dewar

This letter seems to be Dewar's first realisation that his 'direct' liquid air apparatus, built a few months earlier, depended on Joule-Thomson cooling for its efficacy. It is, however, remarkable that in the great burst of controversial correspondence in the scientific and engineering press that followed the publication of Linde, Dewar and Hampson's experiments there was no comment from Kelvin himself. He had, perhaps, lost interest in thermodynamics, to which his last major contribution had been his article on 'Heat' for the 9th edition of *Encyclopaedia Britannica* in 1880, and he was certainly more concerned at that time with Röntgen's announcement of x-rays at the end of 1895 which led to a rapid exchange of 37 letters to and from his friend George Stokes in the first few months of 1896. In May his hand was photographed by x-rays and the print is still preserved at the Royal Society.

Linde's apparatus was on a different scale from those of Dewar and, a little later, of Hampson. He needed liquid air in quantity in order to obtain oxygen from it by the selective evaporation of the nitrogen. His first interchanger consisted of an iron pipe of 10 cm internal diameter down which he had inserted a second pipe of 4 cm internal diameter. The lengths of the pipes were 75 m but the overall size was reduced by coiling them tightly. The assembly weighed over a tonne. The cooling of this mass of metal by gas passing up and down the tubes was clearly a slow business. Thick insulation was provided, but the flow of gas had to be turned off at night when the iron tubes inevitably warmed up a little. It was three days before the interchanger had cooled sufficiently to allow the gas emerging from the expansion valve to liquefy in part. But once this heavy machine was running it produced liquid air on an industrial scale; the first trial yielded eight litres of liquid. He was soon to modify this apparatus and some of its successors had smaller and lighter tubes in the interchanger.

Linde understood exactly what he was doing, for he had learnt his thermodynamics from Rudolf Clausius in Zurich, and he quoted from the work of Joule and Thomson that the cooling of air on expansion is given, in kelvin, by the expression: $[(p_1 - p_2)/4]\,(289/T_1)^2$, where p_1 and T_1 are the inlet pressure, in atmospheres, and the inlet temperature, and where p_2 is the outlet pressure. It follows that the cooling at room temperature and pressure is about 0.25 K for a difference of pressure on expansion of 1 atm. The work done in passing the gas round the apparatus is that of compressing the expanded gas back to the starting pressure; this is proportional to the ratio p_1/p_2, while the cooling is proportional to the difference $(p_1 - p_2)$. The most efficient way of running the plant is to minimise the ratio and maximise the difference, which is achieved by making both pressures as large as possible. Moreover such a choice increases the throughput of gas and so the yield of liquid. Linde chose the upper pressure to be 150 atm and the lower to be 50 atm, thus producing the satisfactory cooling of about 25 K on first passage through the valve. Moreover the cooling increases substantially as the temperature falls below ambient. (Dewar was to choose 150 and 100 atm as his upper and lower pressures.) Linde patented his work in Germany and in Britain,[93] where his provisional application was dated 28 June 1896, a month after Hampson's application.

It seems from the date of his letter to Kelvin that it was the appearance of this report by Schröter that had alerted Dewar to the central importance of the Joule-Thomson effect. Among his papers is an annotated abstract, not in his hand, of Coleman's Friday Evening Discourse of 29 May 1885 in which the annotations mention the Joule-Thomson effect in a discussion of the work of Hampson, Linde and Dewar, but these annotations date from after 1892 (when Thomson became Lord Kelvin) and are probably from 1896.[94] When Joule had died in 1889 Thomson agreed to give a Friday Evening Discourse on his life and work. Unfortunately he had to withdraw from this offer, because of 'pressing business in Scotland', and it fell to Dewar to give the talk instead. It is not clear whether Thomson had any input into the Discourse but it seems that the words were Dewar's. On Joule and Thomson's work he said:

> Its chief object was to prove that in compressing a gas the amount of heat produced is equivalent to the work done, and independent of the mere fact of the approach of the particles. ... They proved that any difference of temperature of the gas brought about by its passage through the porous plug must be due to the work done by it, and that the difference of temperature varied for different gases, according to their constitution. Working under similar conditions, hydrogen was shown to be reduced by a small amount in temperature, air somewhat more (about 0.3°), and carbonic acid a much greater amount ... for instead of the product and the volume being strictly proportional to the absolute temperature, as it had hitherto been believed to be, they found that a new term was involved.[95]

There is much to criticise in this account. The first sentence implies that they were trying to show that the energy of a gas is independent of its density, whereas Thomson thought that this hypothesis was not true and was trying to establish the degree of its falsity. The statement that hydrogen is cooled, but by a small amount, is wrong; Joule and Thomson had shown that it is heated. Moreover it was, by the 1850s, already well-established that for any real gas the product pV was not 'strictly proportional' to the absolute temperature.

The appearance of Schröter's report put Dewar in a tricky position. He had not yet revealed his new 'direct' liquid air apparatus to the public and he clearly wanted to claim the credit for it. Was he to try to acquire as much information as he could about his novel use of the Joule-Thomson effect, or was he to claim that the effect was well-known and so its use was not a true novelty? In the event, he privately followed the first line, while in public he followed the second. To find out more about the Joule-Thomson effect he went to his usual source of advice on problems of theoretical physics, his brother-in-law, Hamilton Dickson. He was a mathematician who had trained at Glasgow under Thomson but had then worked on electrical and mechanical problems mainly related to the laying of the transatlantic telegraph cable. Thomson recommended that Dickson went then to Peterhouse at Cambridge, Thomson's old college, to improve his knowledge of mathematics and physics. This he did and graduated as 5th Wrangler in 1870. Four years later he was elected a Fellow, and he remained at Peterhouse for his whole career, becoming the senior Fellow on the death of Thomson, now Lord Kelvin, in 1907. In about 1878 or 1879 he had married Isabella Catherine Banks, a younger sister of Helen Rose Dewar. He at once began to help his brother-in-law with theoretical difficulties; what was possibly his first letter, dated 20 August 1879, opened in the formal style of 'My dear Professor'.[96] It was not long before this address was replaced by the less formal 'My dear Dewar', but that was as close to familiarity as he ever came, in spite of the family relationship which seems to have been close and friendly; Helen Dewar was 'Aunt Nellie' to the Dickson children. Dickson's extensive contributions to Dewar's work would today have been recognised by joint authorship of some of the resulting papers, but that was not always the custom of the time, and he himself may have wished to stay in the background without any public acknowledgement. His advice to Dewar on how to calculate the cooling by the Joule-Thomson effect was in a letter of 11 December 1895,[97] shortly after Dewar's demonstration before the Managers and the report on it in *The Times*,[83] and only eight days before Dewar was to give a paper to the Chemical Society on his new work. Meanwhile the editor of *The Engineer* took him to task on 6 December for his failure to credit the new apparatus to Linde in his demonstration on 2 December and in *The Times* report on 3 December:

> We have looked in vain in the columns of the *Times* up to the present for Professor Dewar's acknowledgement of his indebtedness to Professor Linde, whose discoveries were based, or were the outcome of experimental study based on some writings of Joule and J.J. Thomson [*sic*].[98]

This protest misses the point, since Dewar could not have had time to design, build and test his new 'direct' apparatus after his (presumed) reading of Schröter's report on Linde's work after 4 October and before his presentation to the Managers on 2 December; it was almost certainly Hampson that he had copied.

The question of what he believed about the principle of his new 'direct' apparatus is further confused by the statement in several accounts of his work that the principle was not new but had been used in the experiments at the RI for the last five or six years. This statement is found first in *The Times* report of 3 December 1895 on Dewar's demonstration before the Managers at the RI the previous day, but it was repeated in an editorial in the journal *Engineering* on 21 February 1896, and by Lennox in the course of his disputations with Hampson in 1898.[99] The claim in *The Times* cannot have been invented by the reporter but must have come from Dewar himself. It is, however, not clear what principle is meant. The only innovation that even roughly fits the times mentioned later would seem to be that of the vacuum flask. Dewar rightly claimed the credit for this invention and always stressed its importance in enabling him to carry out his work, but it was an auxiliary piece of apparatus whose use does not touch on the principle of the Joule-Thomson effect that lies behind the work of Linde, Hampson and Dewar. It is only by supposing that the effect was already well-known that Dewar could have justified the extraordinary claim in his Chemical Society paper that:

> Apart, therefore, from important mechanical details, and the conduct of general working, nothing new has been added by any investigator to the principles involved in the construction and use of low-temperature apparatus since the year 1878.[87]

So it appears that Dewar's private position was that he must understand how his new apparatus depended on the Joule-Thomson effect, and that he relied for that on the explanation offered by his brother-in-law as soon as Schröter's account of Linde's work had revealed the true position. His public position, however, was that both the use of interchangers and the Joule-Thomson experiments were long established parts of the accepted corpus of thermodynamics, and that his apparatus was merely a novel conjunction of effects already well-known. He never claimed in public to have invented any part of his new apparatus. Figure 8.2 shows the suspicious similarity between Hampson's and Dewar's form of the apparatus; Linde's was quite different in form although identical in principle.

There seems to have been some argument between Hampson and Dewar before matters were brought out into the public gaze at Dewar's reading of his Chemical Society paper on 19 December 1895. This was given at one of the regular weekly meetings of the Society at which it was originally proposed that three papers would be read. The newly-elected President, Vernon Harcourt, was in the chair and he was a leader of the recently developed sub-discipline of physical chemistry. It may have been his decision to allow the meeting to hear only Dewar's novel and important physical communication, which was read in full and discussed, whilst six other routine papers in organic chemistry were merely tabled. Dewar described his new apparatus, gave a demonstration of its working and, unusually for him, showed a detailed drawing of its interior. He made no acknowledgement of Linde's prior work, nor of Hampson's patent. His only mention of the origin of the cooling was a vague statement that 'The experiments of Joule and Thomson and Regnault on the temperature of gas jets issuing under low pressures are well-known'.

In the subsequent discussion of the paper his two friends and supporters, Lyon Playfair and Henry Armstrong, condemned the 'undesirable controversy [that] had been going on as to how much originality there is in the researches of Professor Dewar', and declared that 'the attacks made on Professor Dewar during the last few months were disgraceful'. But no particulars of these attacks were made public. The presence of Playfair is particularly significant since he was busy at the time with advising the Prime Minister, Lord Salisbury, on a boundary dispute in South America between Britain and the United States. On the day before Dewar's lecture, President Cleveland had told Congress that if Britain did not back down there might be war between the two countries. The matter was eventualy settled by arbitration but Playfair's presence at the lecture shows how seriously he took his wish to support Dewar. More to the point, however, was the contribution of Bertram Blount,[100] an 'engineering chemist' and a man noted for the vigour of his participation in discussions, who drew the attention of the audience to Schröter's article in *The Engineer* in October which he assumed must have contributed to Dewar's success. William Ramsay told the meeting that Olszewski had already prepared liquid hydrogen in bulk, a claim that Dewar could not refute but on which he cast much doubt. He was right; Olszewski had made no such claim and Ramsay was later to withdraw his remark. Dewar did not deal directly with Linde's claim to priority saying that if his own apparatus had any 'resemblance to the Linde apparatus described by Mr. Blount, chemists, for once, may be congratulated that a small laboratory apparatus works in some respects better than a large industrial

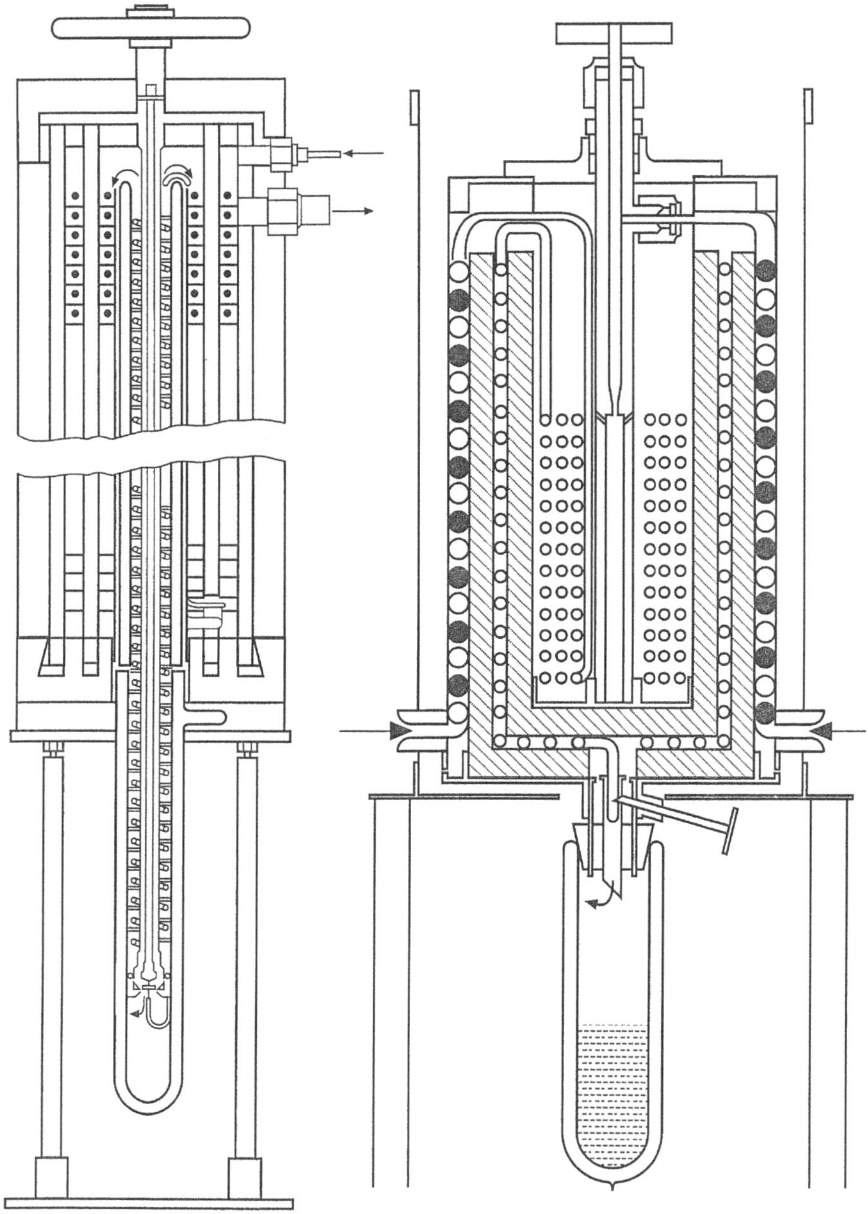

Figure 8.2 A comparison of Hampson's apparatus (left) with that of Dewar (right). The upper part is in each case formed of the coils of the interchanger. In Dewar's apparatus there is also a coil, shown in black, for preliminary cooling with liquid carbon dioxide. In both cases the liquid air comes out at a valve at the base.

plant. … It is a mistake to attribute to Linde the idea of using the "cumulative withdrawal of heat" [Blount's words], for the first time'. After reading these words the editor of *The Engineer* acknowledged on 27 December that 'Professor Dewar, though in a somewhat stinted manner, accorded to Professor Linde the merit of having first thought out and produced the air liquefying apparatus to which attention was called in a *Times* article of December 3rd'. The editor returned to the subject the next week, writing:

> Thus at a stroke, the need for expensive and complicated apparatus designed for liquefying and evaporating various intermediary gases eg. carbon dioxide, ethylene and the like, is removed and the *raison d'être* of such elaborate mechanism as that of which the Royal Institution has hitherto been justly proud, disappears. No doubt that enterprising corporation will wisely replace its obsolete apparatus with a modern and rational plant.[101]

This prediction was not fulfilled since Dewar's next objective, the liquefaction of hydrogen, required its preliminary cooling below its inversion temperature and so the need for auxiliary cooling circuits and elaborate pumps remained. The editor also tried to help his readers with an article commissioned from someone writing under the pseudonym 'X' which was intended to clear up the difficulty of understanding why the new apparatus worked. X did not succeed; he wrote that Schröter:

> seems to hold that the cooling is a result of what has been erroneously called 'Joule's effect' … . But as a matter of fact, the apparatus would work with a perfect gas, and is partly independent of Joule's effect. No doubt that effect [i.e. Joule-Thomson] helps, but it is not in any way essential.[102]

Meanwhile Hampson pressed forward with the building and demonstration of his own apparatus. This was shown 'in the presence of a considerable number of representative men' on Saturday 21 March 1896 at Brin's works in Westminster.[103] His apparatus, Figure 8.2, was remarkably similar to that shown by Dewar to the Chemical Society three months earlier. The main difference was that Dewar had used liquid carbon dioxide as a preliminary coolant while Hampson took air at ambient temperature straight into his interchanger. He was later to lay considerable emphasis on this difference but it is not important since the preliminary cooling serves only to speed up the appearance of the first drops of liquid, and Dewar soon found that he also could dispense with it. Hampson's apparatus ran for 30 minutes before any liquid appeared and then produced it at the rate of about 2 cm^3 a minute. Hampson also said little about why his apparatus worked and left at least one of his audience with the impression that he did not know. The pseudonymous 'Zero' wrote:

> The point to which I wish to draw attention is that no doubt Professor Dewar and others were quite familiar with the facts on which Dr. Hampson has worked,

> but they did not believe that the facts could be utilised to produce liquid gas. In one word, the Hampson process or Linde process, whichever it may be called, apparently is opposed to the teachings of thermodynamics. It ought not to work. The more carefully a man has been taught thermodynamics, the more likely he is to regard the process as impossible; and it is not certain Dr. Hampson, to say nothing of those present at his experiments, understand why the process works. ... Joule found that when air is allowed to expand slowly through a small orifice it is slightly cooled, when hydrogen is so treated it is slightly heated. Scientists availed themselves of this fact, or assumed fact – for the experimental investigation is extremely delicate, and very open to error – and said that this is the reason the Linde process works. As a matter of fact, however, the Joule[-Thomson] effect is quite unnecessary, while its amount is so insignificant that months if not years would be expended in making a quart of liquid air.[104]

Zero's own attempt at an explanation of how the laws of thermodynamics had apparently been thwarted was a vague argument based on the redistribution of the energy of the flowing gas in different parts of the circuit.

Zero was contradicted by Cryos but backed by the article from X and by one from Col. Basevi, a self-appointed authority on the kinetic theory of gases whose own ideas were often wrong, and who wrote a long article doubting the efficacy of the Joule-Thomson effect; he also backed X.[105] Meanwhile Zero and Cryos continued their private battle until the correspondence in this journal was brought to a halt by a definitive article from Linde himself explaining how his apparatus, and so, by implication, how Dewar's apparatus worked.[106] This paper was illustrated by a sketch of the layout of one of Linde's machines which shows how different the interchanger was from those of Hampson and Dewar.

It seems that Zero was correct on one point; there is no evidence that Hampson understood the principle of his apparatus. His patent application and his demonstration in March 1896 contain nothing on the subject and Zero thought he did not understand what he had done. Linde was later to make the same point when, in 1898, he answered a paper by Ramsay in which Ramsay had pointed out that Hampson's patent application had preceded Linde's. Linde wrote that Hampson 'failed to mention the use of the cooling of the flowing gas by the effect found by Thomson and Joule as the source of the coldness'.[107]

Two and a half years after Dewar's demonstration of his 'direct' apparatus to the Chemical Society, Hampson gave an account of his work to the Society of Chemical Industry on 2 May 1898 with the industrial chemist Rudolph Messel in the chair.[108] In it Hampson repeated the claims that Cailletet had invented the vacuum flask and that Dewar had profitted from Hampson's communications with Lennox in late 1894. Dewar and Lennox were both present. There followed a long and heated discussion in which Messel tried to cool the vehemence of the comments, although as an active member of the RI and a generous contributor to Dewar's research fund he was naturally biased in Dewar's favour. Dewar soon demolished the claim on behalf of Cailletet by asking Hampson if he could cite any paper in which Cailletet himself had put forward a claim. Hampson could not, and no more was heard of Cailletet's vacuum flask. The more serious issue of whether Dewar had profited from Hampson's communications to Lennox remained unresolved. Dewar denied any benefit and asked Hampson to withdraw his claim, but Hampson said he had nothing to withdraw; he could not know what had passed between Lennox and Dewar and so was making no accusation, he only wished to put on record the facts of his visit to Lennox and the subsequent development of Dewar's apparatus. There the matter might have rested had Dewar's spectacular communication to the Royal Society of his liquefaction of hydrogen in bulk not followed 10 days later. This announcement unhappily roused Hampson into action again and he repeated in a letter to *Nature* all the arguments recently rehearsed at the Society of Chemical Industry. Dewar naturally responded and there followed another four months of exchanges between the two men that added nothing new of substance but introduced allegations of bad faith and discourtesy. The correspondence had not contributed to the resolution of the differences and it was possibly allowed to continue for as long as it did because the editor, Norman Lockyer, was no friend of Dewar from the time of their disagreements over spectroscopy some 20 years previously.[109] Meanwhile a third stream of correspondence in the journal *Engineering* ranged over several aspects of the disputes; its main theme was the disagreement between Hampson and Lennox on the relative merits of Linde's and Hampson's machines.

8.5 Liquid Hydrogen

Dewar was working on the problem of the liquefaction of hydrogen, apparently the last of the 'permanent' gases to be tackled, from at least the summer of 1894, before he was distracted by the possibilities of Joule-Thomson cooling six months later, and before the discovery of helium on the Earth. The critical temperature of hydrogen was estimated at 35 K or less, and so there was a gap between the lowest temperatures that could be obtained by evaporating liquid nitrogen under reduced pressure, about 40 K, and the highest temperature at which liquid hydrogen could be expected, even under its critical pressure, and an even greater gap between 40 K and the expected boiling point of hydrogen at atmospheric pressure, say 20 K. Dewar's first proposal, mentioned above, was to bridge this gap with an artificial fluid, namely a mixture of nitrogen and hydrogen, as he had suggested in an article in *The Times* in 1894.[110] The experiment failed to give liquid in bulk, perhaps in part because hydrogen is not sufficiently soluble in liquid nitrogen. But by October 1895 he had learnt of the efficacy of Joule-Thomson cooling, and in his letter to Kelvin that month he recognised that this was the way forward for hydrogen. He demonstrated a spray containing liquid hydrogen in his Chemical Society paper in December. This was a repetition for hydrogen of Cailletet's experiment for liquid oxygen and still left him with the greater problem of making pure liquid hydrogen in bulk. In principle, however, he could now see a solution to his difficulties since Joule-Thomson cooling of hydrogen required only that it be expanded at a temperature below its inversion temperature, which is about 190 K at low pressures and somewhat less at higher pressures. This temperature is well above the critical temperature and so such cooling did not need a 'cascade' of substances to get down to the critical temperature; the inversion temperature is well within the reach of liquid air at its normal boiling-point of 80 K and even further above the temperature of liquid air boiling under reduced pressure, which was the route that Dewar eventually used. His first preliminary results were in the Friday Evening Discourse that was based on his Chemical Society paper of December 1895. Hydrogen was here compressed and cooled to 73 K before being passed into a pre-cooled interchanger. A jet of liquid was formed but could not be contained because of its low density and its disturbance by the rapidly flowing portion of the gas that was not liquefied. When the jet impinged on liquid air or oxygen they formed hard solids like 'avalanche snow'.[87] He was slowly mastering the problems of liquefying hydrogen. He made little use of the work of others although when later he came to publish his results he did acknowledge that the Wróblewski's measurements of the isotherms of hydrogen and the use of van der Waals's law of corresponding states had guided him in estimating the critical temperature and other physical constants.

Before going further, however, there was a diversion while he tackled fluorine. This reactive element is difficult to prepare and handle but its physical properties posed no more difficulties than that of liquefying air; its critical temperature is 144 K, lower than that of oxygen (155 K) but higher than that of nitrogen (126 K). The gas had been prepared first in 1886 by Henri Moissan in Paris by the electrolysis of KHF_2 in liquid HF, and he was later invited to give a Friday Evening Discourse on it at the RI. He came, with his laboratory assistant and 'un petit appareil pour essayer la liquefaction du fluor', on 28 May 1897, and had hoped to be able to produce liquid fluorine during the Discourse. This experiment proved to be too ambitious but it was recorded in the Managers' Minutes that:

> Professor Dewar reported that Professor Moissan remained over the Saturday following his Friday Evening discourse on May 28 and they experimented together on Fluorine for a few hours at very low temperatures, and within about two hours they succeeded in obtaining liquid Fluorine. They found that at these very low temperatures Fluorine did not attack glass or indeed any of the materials with which it usually combines at ordinary temperatures with great violence except hydrocarbons.[111]

The condensation of the gas was effected by cooling it by liquid oxygen boiling under reduced pressure, and its own boiling point was estimated at about -186 °C (87 K, the accepted value is now 85 K). These results were published both by the French Academy and the Chemical Society and were reported at the British Association meeting at Toronto in August 1897.[112] The liquefied fluorine was found to show none of the exuberant chemical reactivity of the gas.

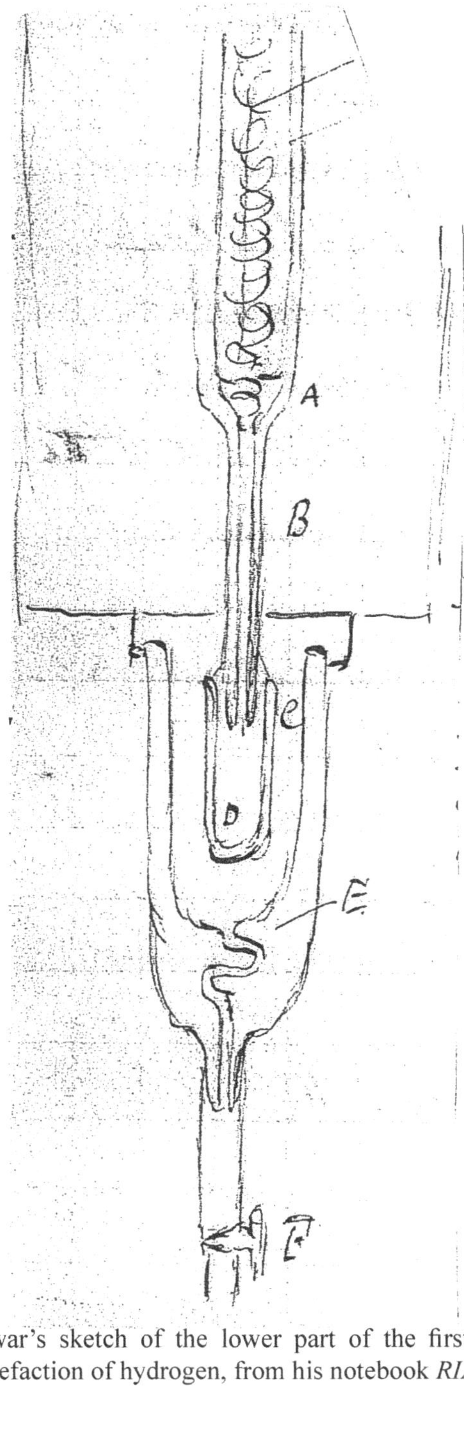

Figure 8.3 Dewar's sketch of the lower part of the first apparatus for the liquefaction of hydrogen, from his notebook *RIA* D V b/1.

Apart from this success with fluorine, the year 1897 was devoted to electrical and magnetic experiments at low temperatures conducted with Ambrose Fleming of University College, and to the designing and building an improved apparatus for his attack on the problem of the liquefaction of hydrogen. There is no adequate description of this new apparatus. None was published and so the apparatus acquired by the Science Museum is that built by Hampson and Travers two years later. The sketch in Dewar's laboratory notebook (Figure 8.3) shows only the arrangement of the unsilvered glass vacuum vessels.[113] This sketch is entitled 'Plan of the Vacuum Tubes for the H[ydrogen] Exp.' and the helical tubes at the top are labelled 'Regenerator H coil'. The legend alongside this sketch reads:

> Silvered Vacuum Vessel with spiral, and opening at bottom A and a long narrow part B so that the end C comes through the aparatus. D is a small vacuum vessel supported in a longer one E having a spiral and opening at the bottom. The top of E is fitted into a brass support that is capable of being clamped by screws and washer to the botttom brass plate of the apparatus. No escape of H[ydrogen] can take place other than into and around D and out through E and stop cock F. It is the regulation of the [amount?] of the escape through F that is so important. It was at the point E on 25th March Exp that the air sucked in liquified. Neither of the v[acuum] tubes D or E have been silvered either wholly or in part as far as the Exps have gone as the chief object was to see what was going on.

An important innovation was the insertion of a helical coil between the two walls of the outer vacuum vessel to take up the strain of the thermal contraction of the glass when cooled to the temperature of liquid hydrogen. He wrote later of this coil:

> This device, developed after many unsucessful attempts at such a vacuum vessel, was found essential for the easy production and collection of liquid hydrogen, and as all the Royal Institution designs for such vessels have been made in Germany, they have been supplied and utilized by other workers unconscious, it may be, where or how they originated.[114]

These vacuum vessels were attached to the bottom plate of the liquefier of which no details are given. It was based on his by now standard 'large' liquefier for air in which carbon dioxide and liquid ethylene were used but here in the final cycle liquid air itself was used, now subject to 'exhaustion' to reduce its vapour pressure to '6 inches of mercury', or 150 mmHg, so producing a temperature of about 57 K. In his Friday Evening Disourse of 20 January 1899, which contains the fullest account of the liquefaction of hydrogen and of the early experiments in which he used it, he recounts that he

> resolved to construct a much larger liquid air plant, and to combine it with circuits and arrangements for the liquefaction of hydrogen. This apparatus took a year to

> build, and many months have been occupied in the testing and preliminary trials. These many failures and defects need not be described.[115]

It was presumably the external appearance of this apparatus that he used to illustrate an article on 'Liquid gases' in the 1902 edition of *Encyclopaedia Britannica* (Figure 8.4). His laboratory notebook gives us a brief narrative of the experiments.[113] The first serious attempt at liquefaction was made on 25 February 1898:

> All apparently went well until after the liquid air was put under exhaustion when attempting to open the H[ydrogen] valve nothing would come. Nothing would remove the obstructions. Thought it was solid air but the stoppage remained hours afterwards so it looks like some solid carried forward at valve. ... It was found that the obstruction was due to solid particles of solder carried by H.

> 14th March/98. Between 25th February and this date had repeated failure in getting Silvered Vacuum Tube to fit the Regenerator ... Started the apparatus as described on the 25th Feb^{y} ... After a few minutes a white solid along with a [?] of liquid, say 5 cc, collected in the inner of the two vacuum test tubes round which the H coming from the Regenerator [was] passing.

> 25th March/98. ... Since the previous Experiments we placed an enlarged coil of [?] pipe of 600 cc capacity over the end of the H. coil in the L[iquid] H[ydrogen] chamber before it goes [through] the Regenerator in order to catch liquid air and other impurities in the H. The whole arrangement seemed to work well.

> 25th April. Repeated the above experiment ... 10th May/98. Started Hydrogen apparatus treating CO_2, C_2H_4 and air circuits as in previous Exp^{t}. This time the liquid was collected for 2½ hours and the air pump put on for the last half hour. ... After the ½ hour the air supply was stopped and the air pump was allowed to get the pressure reduced to 6 inches of mercury. The H started running through the valve at 175 atmospheres and about the rate of 15 c. ft a minute [7 litres a second]. Shortly after starting the nozzle plugged but it got free by good luck and almost immediately drops of liquid began to fall from the V[acuum] tube into the [inner ?] V [tube] and soon accumulated to 20 cc in volume. There was a great escape of H through the [stopcock] and the outer Vacuum Vessel was covered with ice. The H was a clear transparent liquid with a well defined meniscus (even better than liquid air) showing no absorption spectrum and as long as the surroundings of the [inner] vessel were cool seemed to evaporate very slowly. I now took off the [vacuum] vessel from the apparatus and placed in the liquid H a narrow quill tube; sealed at the end immersed and open at the other; when immediately solid air appeared in tube. In the same way the He[lium] tube was placed in the liquid H and I could see that a liquid was formed. This He tube gave nothing when placed in liquid air under exhaustion. The H vapour above

> the liquid had a very marked appearance and seemed very dense with circulating currents of a greenish blue cloud. … The yield of H in the liquid is about 1/6 that of liquid air or say about 1%. As the liquid air was boiling under 6 in. pressure as the H at starting was at -204 °C. The H gas before use was tested for the presence of air and was found not to exceed 1%.

By chance Rayleigh was present in the RI when hydrogen was first liquefied. That day, a Tuesday, Dewar asked the President, Lord Lister, if he might announce his success at the regular meeting of the Royal Society on the Thursday. He and Fleming had already contributed a paper for that day's meeting. He continued, however, with further experiments on 12 May until he had to leave off his work to go to the Society's meeting in Burlington House.[115] His notebook continued:

> 12 May Conducted everything as on the 10th starting with H gas at 140^{atm} and [?] collected about 60 cc and took it out of the apparatus placing the V[acuum] tube in liquid air so that the outer surface was always [?] -183°. … Dipping some cotton wool into the liquid and lifting into the air it ignited with a [?] giving a [?] of flame. The cotton wool saturated with L[iquid] H[ydrogen] between the poles of a magnet was attracted. Believe this due to liquid air being condensed on its surface as solid.

> From 19th May to 27th May had a succession of failures due to the breaking of the vacuum vessels and the necessity of making a new coil and vessels to fit. On 27th May got a quantity of H as liquid about 30cc and then no more due to usual stoppage of coil.

Dewar announced his success to the Royal Society on 12 May and again to the Chemical Society a week later. The claim to have liquefied helium was quietly abandoned after its mention in the titles of his papers to the Royal Society and the Chemical Society. He offered no explanation for dropping the claim, but he had presumably been misled, as were others before him in similar cases, by some impurity in the hydrogen, perhaps neon.[115]

The liquefaction of hydrogen was a fitting climax to Dewar's cryogenic work. There were the expected announcements in the press and presumably private letters from his fellow scientists although he seems to have preserved none of these. That from Kamerlingh Onnes must have been the most prized and led to a reply from Dewar on 11 June:

> It is very kind of you to speak so kindly about hydrogen. You know something about the difficulties of the problem. My troubles I can see are only beginning. It will be a long time before hydrogen is *on tap*. No doubt it will come some day. In the meantime progress must be very slow. I enclose an account of the BP [boiling point] and the Density determinations as far as I have been able to do this.[116]

Figure 8.4 Engraving of the apparatus for liquefying hydrogen, from Dewar's article on 'Liquid gases' in the 10th edition of *Encyclopaedia Britannica* of 1902.

He next solidified the hydrogen in 1899 by cooling the liquid by evaporation. He estimated the temperature of the triple (or melting) point by gas thermometry to be about 16 K, a little above the currently accepted value of 14 K. By pumping on the solid he cooled it further to what he believed was about 13 K, but may have been as low as 10 to 11 K.[117] The centenary of the RI fell in 1899 and was marked by lectures from the two professors, Rayleigh on the work of Thomas Young, and Dewar, naturally, on liquid hydrogen.[118] The years 1898 to 1901 were occupied with further experiments on the physical properties of liquid hydrogen,[118] on spectroscopy at low temperatures (with George Liveing),[119] and on thermometry,[120] only briefly hindered by another explosion on 12 December 1898.[121]

The experiments on hydrogen had led Dewar into greater expense than any of his earlier work. He had previously obtained some support from the Goldsmiths Company and from the Royal Society but much of the money came from his own pocket and from those of the keenest and wealthiest of the members of the RI. After his first success with the Joule-Thomson method in December 1895 there was a contribution of £100 from *The Times*, probably instigated by their correspondent Munro Ross. It was estimated that from 1888, the year he took over from Tyndall as Superintendent of the House, until his death in 1923, Dewar had himself contributed £2500 to the Fund for Experimental Research.[122] He also gave the £69 prize money from his Rumford Medal from the Royal Society. He enlisted help from his relatives also; his brother Hugh in Australia contributed several times, as did Charles Scott Dickson, a barrister and sometime MP, a brother to Hamilton Dickson and also brother-in-law to James Dewar. Among the many members of the RI who supported his work the most prominent were Ludwig Mond (who probably contributed about £3500 in all, starting with an annual donation from 4 July 1887),[123] Frederick Bramwell, Rudolph Messel, Frederick Abel, Andrew Noble, Lachlan Rate (a Visitor) and a Mrs Bromfield Moore. Mrs Moore was an American theosophist who contributed a total of £480 on her visits to Britain. In 1888 she had offered also a cheque for $5000 but this was later returned when it was found that she had attached some unacceptable conditions. In 1899 there was a letter in *Nature* that claimed that the money was connected with her support for an American inventor, John Keeley, who was trying to interest the world in a new vibratory force and that Dewar had agreed to support his claim in return for this cheque. Dewar had, in fact, done nothing of the kind and had seen at once that Keeley was a crank. The RI had returned the money as soon as the purpose of the donation became clear.[124] The total cost of Dewar's experiments on liquefaction was probably about £16000.

Some years later there came an opportunity to build a larger and better-engineered apparatus for liquefying hydrogen. An international exhibition was to be held at St Louis, Missouri, in 1904 to which all nations were invited to show what they could do in scientific, technical and artistic achievements. The British entry was to include an apparatus for the production of liquid hydrogen, an operation not yet attempted in the United States. The apparatus was ordered from Lennox, Benton and Reynolds but this time the Royal Commissioners were paying.

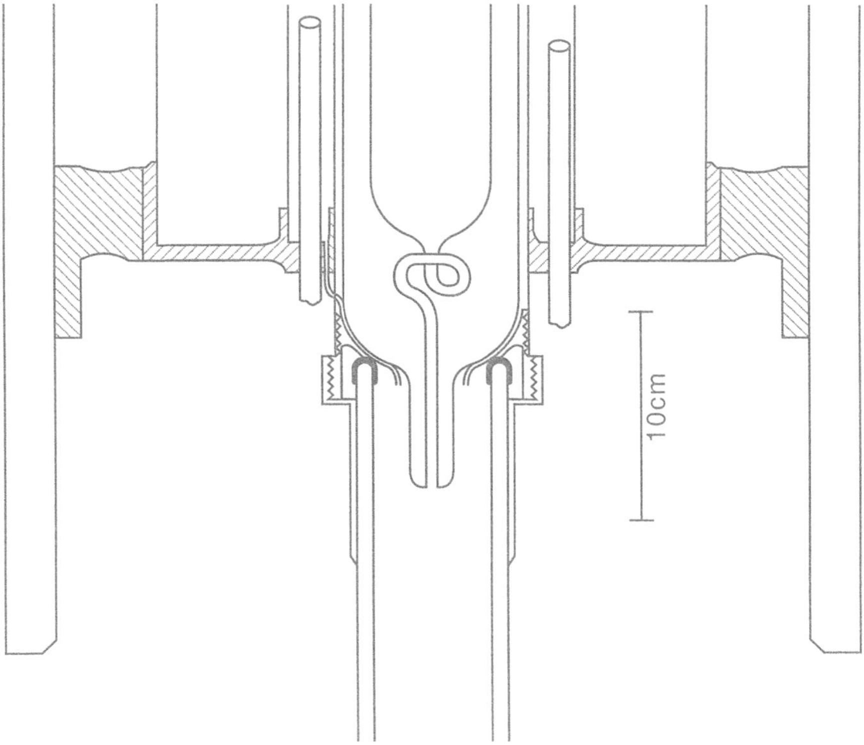

Figure 8.5 Drawing from the blueprint of the apparatus built in 1903 for the St Louis Exhibition. The central tube at the top contains the liquid hydrogen which is surrounded by the vacuum flask. The liquid is delivered into the lower detachable vacuum tube through the helical capillary.

Before shipment to America, the equipment had to be approved by the oil engineer Boverton Redwood who was the chairman of the sub-committee organising the British entries in the chemical section. The apparatus was built towards the end of 1903 when Redwood wrote to Dewar to say the testing could best be done under Dewar's supervision at the RI, to which Dewar agreed as long as the RI was not held responsible for any damage that might occur.[125] There was then some dispute that led Dewar to withdraw his cooperation and his offer to supply the vacuum flasks with helical capillaries that could be made then only in Berlin. Fortunately Robert Hudson, one of the British engineers involved, managed to get a fresh supply directly from Berlin which arrived in St Louis just in time.[126]

The man put in charge of the operation of the equipment in St Louis and with explaining its operation to visitors was Joseph Petavel[127] who had worked in the Davy-Faraday Laboratory. There is no indication that Dewar himself ever intended to go to St Louis and, in the event, his health was not good in the summer of 1904 and this circumstance alone would have prevented him demonstrating his plant in person. This version of the hydrogen apparatus was close in design to the original of 1898, and the RI has a partial set of blueprints of the core of the apparatus and of the pumps (Figure 8.5). This apparatus must be considered the finest achievement of Lennox's company. It produced liquid hydrogen twice a week for the two months of the exhibition without a single hitch. Great care was taken with the purity of the hydrogen which was generated from zinc and sulphuric acid. One part in a 1000 of any impurity would block the jet through which the gas was expanded. Each public demonstration by Petavel required between one and three litres of liquid hydrogen, amounts that were then quite unprecedented. Some of this hydrogen was solidified and also shown to the visitors. Around this spectacular achievement Petavel gave over 80 lectures for the public at large and for professional physicists, chemists and engineers. At the end of the exhibition the apparatus and its demonstration received a Grand Prize, the highest award, and Dewar and Petavel received respectively a gold and a silver medal. The equipment never came back to Britain but was bought by the US Government, presented to the National Bureau of Standards, and re-erected in Washington.

8.6 Cryogenic Experiments

Dewar always maintained that the liquefaction of air and of hydrogen were only means to an end; what he was really interested in was using these coolants to investigate the properties of matter at low temperatures. He maintained correctly that the closer one got to the absolute zero of temperature the more the temperature scale 'opened out';[128] that is, roughly speaking, there was as much new physics to be found between 1 and 10 K as there is between 10 and 100 K. He did not go as far as modern theoretical physicists who prefer to work in a scale of β, which is a reciprocal energy, $1/kT$, where k is Boltzmann's constant, but he might have shared their idea that such a scale is more enlightening than the the conventional 'absolute' or Kelvin scale. The 'absolute zero' of the usual scale, $T = 0$, is at $\beta = \infty$, a limit that perhaps suggests more strongly that the state is unattainable. His success in liquefying hydrogen therefore opened up a further range of temperature for new investigations. But what experiments should he make? It was here that Dewar's lack of appreciation of current physical theory let him down.

By the end of the nineteenth century the theoretical structure that we now recognise as classical physics was essentially complete. It rested on Hamilton's mechanics, Gibbs's thermodynamics, Boltzmann's statistical mechanics, and Maxwell's electricity and magnetism with their link to optics. No sooner was the structure thought to be approaching completion, however, than cracks began to appear – cracks that were to defy attempts to cure them within the structure itself. At the very end of the century Kelvin gave a Friday Evening Discourse in the RI, which Dewar presumably attended, and in which Kelvin drew attention to two of these problems; his title was 'Nineteenth century clouds over the dynamical theory of heat and light'.[129] He criticised the classical theory of the equipartition of energy and its failure to describe correctly the heat capacities of the simple gases. A few years later he was baffled by what seemed to be a violation of the first law of thermodynamics by the apparently inexhaustible source of heat contained in the newly discovered radioactive elements. Other failures of classical theory were its inability to handle the energy distribution of black-body radiation and the photoelectric effect. Atomic spectroscopy presented a long-standing problem that Liveing and Dewar clearly knew about but which they never seriously discussed. Spectroscopy had been for 50 years a useful analytical tool for the chemists and was then proving its value in the characterising of the newly discovered inert gases, but how such complicated effects as line spectra could be reconciled with the presumed simple structure of atoms was a mystery, as was also the bonding of atoms to form the chemists' molecules. Other branches of experimental physics were yielding new, exciting and so far inexplicable facts, with the discovery of X-rays, radioactivity and of the sub-atomic particle that came to be called the electron. Dewar was as keen as any man in hearing about the new experiments but he showed no interest in the theoretical problems to which they gave rise and this lack of curiosity led to his own work at low temperatures making less impact than it might have had. Planck's and Einstein's quantum theory seem to pass him by, except for a brief reference to Einstein's theory of the heat capacity of solids (see below, pp. 126–7). Nernst's heat theorem of 1904, or the third law of thermodynamics as it soon came to be known, might be thought to be central to his interests in the behaviour of matter as the 'absolute zero' was approached, but it seems that he never mentioned it in his papers. This weakness in his interests became apparent when he addressed the British Association as its President in 1902, and is discussed again in Chapter 11.

One aspect of work at low temperature that had theoretical implications was the measurement of temperature itself. The first Conférence Générale des Poids et Mesures in 1889 adopted the constant-volume hydrogen thermometer as the standard scale. It was a good choice from a practical point of view since hydrogen remains a gas at (what were then) all accessible temperatures and, moreover, it has unusually small deviations from the perfect-gas laws. It was a surprisingly long time before the practical realisation of the thermodynamic temperature was achieved by extrapolating a temperature on a gas scale to zero pressure. Nowadays we accept without argument that the product pV departs from its limiting value of RT by a leading term that is linear in the density, or V^{-1}, but such a simple result was not clear to Regnault, for example, in the middle of the century, and so the departure of a real gas scale from the perfect gas scale, which is also the absolute thermodynamic scale, was something that required special study. It was this departure that Joule and Thomson attempted to calculate in the 1850s. Their experiments were accurate but their calculation was at fault, as was first pointed out by H.A. Rowland in 1880. Fortunately the difference between the hydrogen scale and the thermodynamic scale was sufficiently small not to affect the accuracy that

Dewar and his contemporaries were aiming at. Today the correction is more easily calculated by measuring the pressure directly rather than going through the Joule-Thomson effect, but such measurement of isotherms never formed part of Dewar's programme. For practical use, however, a gas thermometer is bulky and difficult to handle and a more useful measure is that of a metallic resistance thermometer. The relation between the platinum resistance thermometer and the gas scale was part of the subject of several of Dewar's papers in which he was helped by Hamilton Dickson. One paper by Dickson himself was so closely related to Dewar's use of the resistance thermometer that it was reprinted as an appendix to the second volume of Dewar's *Collected Papers*. Another paper, by Dewar's collaborator Ambrose Fleming, which was also included in the same volume, gives a concise account of his and Dewar's use of the platinum resistance thermometer and its relation to the thermodynamic scale.[130]

A third form of thermometer that is convenient to use at low temperatures since its size is small enough for it to be accommodated in most cryogenic apparatus, is the thermocouple. If three wires of two different metals are connected in series, as for, example, iron–lead–iron, then there is a difference of voltage across the two outside ends of iron which is proportional to the difference of temperature of the two iron–lead junctions. The theory of such differences was then quite obscure and so the voltage from each pair of metals needed calibration against, ultimately, the standard gas thermometer. Dewar's first attempt to measure the boiling-point of hydrogen with a thermocouple gave what were clearly misleading readings and he had to go back to the hydrogen-gas thermometer. Dickson compiled a long list of calibrations for lead against many other metals from 73 K to 373 K, which was again included in the second volume of Dewar's collected papers.[131]

It was not until 1911 that the Physikalisch-Technische Reichsanstalt in Berlin proposed that the absolute scale be adopted formally as the international scale of temperature and this suggestion was not fully realised until the magnitude of the unit of temperature (the kelvin) was fixed by defining the triple-point of water, where solid, liquid and gas coexist, as 273.16 K. This choice, made in 1968, ensures, as closely as possible, that there is an interval of 100 K between the freezing point and the boiling point of water, both at atmospheric pressure.

The cryogenic problems that Dewar tackled at length arose mainly from his collaboration with two colleagues, the long series of spectroscopic experiments made with George Liveing at Cambridge and the electric and magnetic experiments made with Ambrose Fleming at University College, London.[130] The first cryogenic work in spectroscopy followed soon after Dewar's success in liquefying oxygen and nitrogen and consisted only in a study of their absorption spectra and some not very precise measurements of the refractive indices of the liquids. Liveing and Dewar were, however, surprised by the similarity of the spectrum of oxygen as a liquid and as a compressed gas, a similarity which 'implies a persistence of molecular constitution which we should hardly have expected'.[132] When liquid hydrogen became available then more ambitious experiments could be undertaken, both on the spectra of helium and the other newly discovered inert gases in the atmosphere, and on the use of liquid hydrogen to condense out all gases other than helium and so produce extremely high vacua. This work included the use of charcoal cooled by liquid air or hydrogen to achieve high vacua in both glass and metal vessels.[133]

Dewar's interest in magnetic properties may date from his dramatic observation of the behaviour of liquid oxygen in the field of Faraday's great electromagnet. On 10 December 1891 he wrote to Thomson, then the President of the Royal Society, to describe his observation and to suggest that Thomson might care to announce it that afternoon at the regular meeting of the Society. He wrote:

> to my surprise I have witnessed the liquid oxygen, as soon as the magnet was stimulated, *suddenly leap up to the poles and remain there permanently attached until it all evaporated.* To see liquid oxygen suddenly attracted by the magnet is a very beautiful confirmation our knowledge of the properties of gaseous oxygen.

A week later he wrote again to say that liquid ozone behaved in the same way. He could not resist making such dramatic observations the subject of a Friday Evening Discourse.[134]

Dewar's work with Fleming covered observations that were related to the establishment of temperature scales, as mentioned above, and also other experiments that ranged more widely. The collaboration was described by Fleming in a letter to Henry Armstrong written after Dewar's death:

> He had a charming and hospitable side but was not easy to work *with* unless one gave way continuously to his rather dominant personality. His chief characteristic in scientific work seemed to be the *thoroughness* with which he probed a complex matter to its bottom & his hatred of any shams or superficialities. Like all forceful characters, I think that people either liked him very much or the opposite. He certainly grappled with enormous skill with very difficult problems such as the liquefaction of hydrogen, and had, like all of us, a great battle for getting the £.s.d. for doing anything. As far as the electric & magnetic work that we did together it ought to be noted that all the 20 or so papers we published were all actually written by me although discussed between us before publication and all that part of the work which did not require low temperatures was done at University College. The work in fact grew out of a suggestion I made to him in or about 1892 that it would be interesting to measure the resistivity of carbon filaments in liquid oxygen and see if the resistance increased. The first paper published by us in 1892 was preliminary and the difficulty of keeping constant the temperature turned his thoughts to the vacuum vessel ... [Fleming's account of the making of the first vacuum vessels is set out above in Section 8.2.].[135]

The electrical and magnetic measurements covered a wide range of inorganic materials, both metals and non-metals, organic substances, and ionic solutions.[136] The results, as Fleming confessed, were a useful compilation that might serve future theoretical physicists as a foundation of their work but contributed little to their current understanding.[137] The most important generalisation, which they established as soon as they obtained their first preliminary results, was that for pure metals the resistivity seemed to approach zero as the temperature went to zero. Alloys, or metals with only traces of other elements, behaved quite differently; the resistivity generally fell on cooling, but the apparent intercept at zero temperature was substantial. Their experiments were made before liquid hydrogen, let alone liquid helium, became available in 1898 and 1908 respectively, and so they were a long way from discovering the sudden fall to zero resistivity at a non-zero temperature – that is, the phenomenon of 'supraconductivity', or 'superconductivity' as it is now called. They did, however, make some careful experiments on liquid and solid mercury, perhaps the easiest metal to purify to a high standard, with which Kamerlingh Onnes discovered the transition to the superconducting state at 4.2 K in 1911.

Parallel with this fact-finding attack on the electrical and magnetic properties of the metals was a long study of the specific heats of 53 solid elements. This was undertaken in collaboration with William Green, but, since Green was an assistant, it was published by Dewar alone. He had had a long interest in calorimetry and made a series of measurements at liquid nitrogen and liquid hydrogen temperatures. His calorimeters were based on measuring the volume of gas evolved when a sample was dropped into the cryogenic liquid. He found that although most solid elements had similar specific heats at room temperature they were smaller and differed widely at low temperatures. When the specific heats of the elements at low temperatures were plotted against the atomic number the line showed a series of peaks that followed the Periodic Table, in a way that was similar to that of their atomic volumes, a periodicity which had been studied previously by Lothar Meyer. We can see now that the change from similarity at room temperature to smaller values and periodic change at low temperatures is a quantum effect that could be interpreted in terms of Einstein's theory of specific heats, but at the end of the day Dewar ducked any attempt at an explanation, sheltering behind the excuse that 'until more acurate values' are available 'it will be advisable to delay theoretical discussion'. That day never came.[137] Einstein's first paper on the quantum theory of the specific heats had appeared in 1907 but it was a few years before Nernst and others drew attention to its importance in physical chemistry through his 'heat theorem'.[138] By 1913, however, the fall of specific heats to zero at the absolute zero of temperature had been discussed at the Solvay Conference of 1911 and elsewhere. Dewar never considered such quantum effects in his published work; the nearest he came was the inclusion of a graph of Einstein's specific heat in a paper of 1912, but his rather casual dismissal of it as something that might be explained by some 'development of electron theory' suggests that he had not troubled to attempt to see Einstein's foundation of his result in the quantisation of energy. It is, however, remarkable that he had recorded the graph at all at a time when few appreciated its importance.[138]

The discovery of radioactivity and, in particular, the preparation of radium on which Pierre Curie lectured at the RI on 19 June 1903,[139] had already led Dewar into this fashionable field. In March Curie had shown that radium emitted heat in an apparently continuous flow. Dewar and Crookes showed, in a paper read read to the Royal Society on 28 May, that this flow of heat continued without any diminution even at the temperature of liquid hydrogen.[140] He then collaborated with Curie to show that the gas emitted from a radium salt was helium, and was later to determine the rate of this emission.[141] This rate is difficult to measure since the volume of helium produced is exceedingly small. The isotope 226 of radium is an alpha-emitter and has a half-life of 1600 years, so 1 g of radium would produce about 0.13 mm^3 of helium at atmospheric temperature and pressure in a day. But the product of this emission is radon, itself an alpha-emitter with a half-life of only 3.8 days, and the decay of this leads to further alpha and beta emitters, so the total emision expected from radium in equilibrium with its daughter isotopes is closer to 0.4 mm^3 a day. Dewar first found 0.37 mm^3 but after further experiments and a recalculation arrived at 0.50 as his best estimate. He noted, probably with pleasure, that Ramsay had found a result about twice his, but both were unsure about the possible liberation of hydrogen from the action of the radiation on any traces of water or organic contaminants.

He studied also many phenomena that were well-known at ambient temperatures to see how they were changed by cooling to the temperature of liquid air or liquid hydrogen. These included the rates of fast chemical reactions, the activity of photographic plates, phosphorescence, and the viability of bacteria. Nothing unexpected was found.

Chapter 9
Argon and Helium

We have seen that in 1882 Lord Rayleigh, as President of Section A of the meeting of the British Association for that year, had said that he proposed to look further into the validity of Prout's hypothesis that atomic weights were integral multiples of that of hydrogen. To do this he resolved to re-measure, with as great an accuracy as he could, the densities of hydrogen and oxygen, and see if their ratio is an integer. Some years later[1] he obtained the result that it is close to an integer but departs from it by an amount that was outside his experimental error, namely 15.882. Not content with this result, he decided to measure also the density of nitrogen. There are two obvious ways of preparing pure nitrogen. The first is to start from air, dry it, remove any carbon dioxide, and then separate the nitrogen and oxygen by removing the latter by heating the air in contact with copper to form cupric oxide. The second is to take a nitrogeneous compound that can be purified chemically and extract the nitrogen from it. Rayleigh chose first the 'atmospheric' route and found that seven weighings were consistent to one part in 10^4. For the second route he chose a method proposed originally by Vernon Harcourt and suggested to Rayleigh by William Ramsay. Air is bubbled through a strong solution of ammonia before passing over hot copper when the hydrogen in the ammonia is oxidised to water and the gas then dried. Only about one part in seven of the nitrogen comes from the ammonia, but by substituting oxygen for air the whole of the resulting nitrogen could be 'chemically' sourced. He found that the two methods disagreed; however careful he was, and whatever variations he tried, the atmospheric nitrogen was always about half a percent heavier than the chemical. This difference was about 50 times the variations found between the weighings and so beyond any experimental error of measurement; something was amiss with the purification achieved in one or both of the routes. Rayleigh was no chemist and had written to *Nature* in September 1892, before the experiments with ammonia and oxygen, to put the problem before the public and to ask chemists for any suggestions for its resolution.[1] Ramsay replied but, as he said himself, he could do no more than describe at some length the facts as they now stood.[2] The more doubtful of the two procedures seemed to be the first; what is the evidence that dry air, freed of carbon dioxide, contains only oxygen and nitrogen? Rayleigh put the question to Dewar, his chemical colleague at the RI, who replied that he knew of nothing being done on the question since Cavendish's experiments in 1785, in which he had sparked air to induce the two components to react, and had then absorbed the oxides of nitrogen in alkali. More oxygen was added and the sparking continued until, it was assumed, all the nitrogen and oxygen had been removed. Cavendish found that there was a small residue, about one part in 120 of the original air, which defied further oxidation. He attached, however, little importance to this residue noting only that if there were any other gas present in air it was no more than 1/120 of the whole.[3] (A question arose later of whether it was Dewar or Ramsay who first told Rayleigh of Cavendish's experiment. The point would not merit discussion were it not for the later arguments that surrounded all the work of Dewar and Ramsay, but the evidence seems to show that Rayleigh first learnt of Cavendish's work from Dewar in April 1894, although Ramsay may also have mentioned it to him shortly afterwards.[4])

Rayleigh followed Dewar's advice to read Cavendish's paper and decided to repeat his experiment. The removal of the nitrogen was a tedious operation but by 6 August 1894 he had about 0.3 cc of a gas that was neither nitrogen nor oxygen, from the spectrum seen during the sparking. Meanwhile Ramsay had himself had entered the field with what was a more efficient way of removing the nitrogen, namely by reacting it with hot magnesium to form the nitride. He started such experiments in April 1894. Such a trespass into another man's field was thought by some to be contrary to scientific etiquette and Ramsay's action was criticised by Kelvin and others.[5] Rayleigh himself seemed not to have actively welcomed it but he recognised that Ramsay's method of producing the hitherto unknown residual gas was working better than the one he had adopted from Cavendish. He had, moreover, in his letter to *Nature*, invited help from the chemists, so perhaps he had little to complain of when one took him at his word. He persisted, however, with his use of Cavendish's route with a de Meritens alternator which he had borrowed from the Cavendish Laboratory in Cambridge and with which he eventually obtained 1 cc of the new gas. Ramsay wrote to him on 4 August 1894 to say that he had now isolated about 100 cc of the new gas which had a density of 19.075 (H = 1; the current figure is 19.032).[6] He had found it not to react with any of the reagents that he had tried. Rayleigh and Ramsay decided to join forces and to present together a paper to the forthcoming meeting of the British Association in Oxford later that month. The paper was read by Rayleigh on 13 August to a joint session of Sections A (Mathematics and Physics) and B (Chemistry) and naturally aroused considerable interest, but also scepticism. Chemists were loath to admit that the atmosphere, which they had studied for centuries, contained 1 per cent of what seemed to be a new element. H.G. Madan, an Oxford chemist, noting its apparent inertness, proposed for it the name 'argon', from the Greek ἀργόν, 'idle'. Among those who doubted that it was a new element was Dewar, who wrote two letters to *The Times* immediately after the meeting at Oxford. On 15 August he wrote that:

> One is forced to the conclusion that if the new nitrogen is present to the extent of ½ to 1 per cent, and has a density of about 19, then it is, indeed, a strange substance, being as volatile as nitrogen or oxygen, and, therefore, not capable of separation by difference of boiling point.[7]

He wrote again the next day to add:

> The inference is that the new substance is being manufactured by the respective experimenters, and not separated, as they imagine, from the ordinary air. If the suggested view is correct it is certain that the allotropic nitrogen would be far less volatile than the ordinary variety, just as the vapour-pressure of red phosphorus is small compared with that of the white, and, therefore, it ought be separated from liquid air by its high boiling point, even if it does not become solid at the existing temperature.[7]

Dewar was conservative in his attitude to unexpected discoveries. When, for example, a few years before the discovery of argon, Nikola Tesla announced that there could be long-range electrical effects without wires, Dewar was dismissive of the claim until Tesla demonstrated a 'Tesla coil' in a Friday Evening Discourse.[8]

Dewar now looked also into the effect of magnesium on 'chemical' nitrogen and found that it too appeared to generate a residue that could not be removed by reaction with the magnesium. By November he had had an opportunity to look into the liquefaction of the two forms of nitrogen. He wrote to Kelvin on 23 November 1894:

> Dear Lord Kelvin,
>
> When I wrote the letter to the Times re the new element I had then no opportunity of looking more carefully into the liquid air side of the question; owing to the apparatus being disbanded. Now I have got things into working order again. Yesterday I compared the physical [behaviour?] of the Nitrogen of the Air with the Nitrogen separated from a chemical compound. Both samples had been carefully treated by the same chemical reagents and were therefore strictly comparable. On liquefaction, I could observe no trace of difference in the appearance of the liquids. This shows that the *solid impurity in 'air'* as ordinarily liquefied has nothing to do with the *New Substance*. I compared the two samples side by side and could observe no difference either during evaporation or condensation. On cooling to -200 °C *nothing separated*. The conclusion appears to be therefore the new element does *not liquefy* or is *dissolved* in the *ordinary Nitrogen* and has approximately the *same boiling point*. Before making any corrections in the Times Letter I will repeat the experiments. …
>
> Yours Very Truly,
>
> James Dewar[9]

He wrote to Rayleigh on 4 December 1894:

Dear Lord Rayleigh,

I am going to give a little paper[10] to the Chemical [Society] on Thursday on the examination of gases at low temperatures. Indirectly this in some points bears on the work you have in hand, and in case of discussion I would like to know the exact facts. From my experiments it is clear that chemical nitrogen treated with magnesium produces a new gaseous product. So that the magnesium method of separating chemical nitrogen from the new constituent breaks down. … In the meantime the point of chief interest is the result of the comparison of your product with that of Ramsay. *Am I right in assuming that your substance is the same as that of the Ramsay product ??* If this is the case, it appears that chemical nitrogen can be transformed into the air product by the action of magnesium, or the air product is dissolved in magnesium or results from its action on glass ?? … I have just learnt that Dunstan has confirmed the action of magnesium on chemical nitrogen.

Yours very truly,

James Dewar

Rayleigh replied rather coolly the next day:

Dear Prof. Dewar,

I have received your letter of yesterday, which places me in rather a difficult position. Prof. Ramsay and I are as you know working in concert, and have published nothing as yet, beyond an oral statement at Oxford in a few sentences of what we had done, and announcing that we were pursuing the subject. Matters are not yet ripe for publication. Under these circumstances it appears to me to be contrary to scientific usage for others to bring forward the questions involved. … I should be glad to tell you what I know privately, but not for the purpose of a public discussion, which in my view ought not to take place. … Would it not be better to wait until we have published our work, and, then, if we make mistakes, as is not unlikely, to correct them?

Yours very truly,

Rayleigh

Dewar quickly backed off, telling Rayleigh that 'I never had any intention of treating your work. My informant Dunstan may be all wrong, and my question was out of mere curiosity; never intending to use the same in public'.[11]

The Chemical Society meeting was held that evening and a report in *The Times* the next day, 7 December, made it clear that strong opinions were expressed, particularly by Armstrong, then the President of the Society, who was in the chair. Rayleigh and Ramsay had decided not to attend. The report shows every sign of having been written by *The Times* correspondent Munro Ross, the friend and often the mouthpiece of Dewar and Armstrong. It read:

> THE NEW ELEMENT
>
> There was an unusually large attendance at the Chemical Society last night, in anticipation of a discussion upon the new element announced by Lord Rayleigh at the meeting of the British Association. … It was therefore a serious disappointment to the Chemical Society last night to discover that not one of the men known to have been engaged in working at the new element came forward to give information as to the results so unequivocally proclaimed. Some astonishment was also felt when the president of the society [Armstrong] observed that a good deal of feeling had been called forth by this question, notwithstanding its purely scientific character. It was evident from his remarks that the discoverers of the new element are anxious to forbid discussion, on the extraordinary ground that as they have not published their conclusions, discussion can only take place upon private and confidential information. It is obvious, as he pointed out, that when chemists are informed that they have entirely failed to comprehend the constitution of a substance upon which they have bestowed so much labour as upon the atmosphere, they have an indefeasible right to carry out whatever experiments they think fit. Moreover it is absurd after not only the facts of the discovery but details of the preparation and properties of the new element have appeared in the public press to pretend that there is any breach of social or scientific etiquette in discussing them. Why the discoverers should impose upon themselves and seek to impose upon others the extraordinary reticence concerning what has been heralded from the highest authority [Kelvin] as the greatest scientific event of the year is more than anyone is prepared to explain. …
>
> Prof. Dewar described last night the methods of applying liquid air to the investigation of the properties of gases. It appears from his experiments that chemically prepared nitrogen liquefies at the same temperature and boils off at the same rate as nitrogen obtained from the atmosphere. Yet according to the discoverers of the new element, one contains a substance which is not present in the other, the density of which is nearly half as great again as that of nitrogen. It follows either that the new substance does not liquefy at all, even at temperatures which condense much rarer gases, or that it behaves in exactly the same way as nitrogen. Chemists will fully appreciate the extreme singularity of a substance with the assigned density which fulfils either condition. It is not too much to say that its discovery would revolutionize chemical theory. But the whole question becomes infinitely more obscure if, as seems to be the case, chemically prepared nitrogen passed over red-hot magnesium behaved in a manner indistinguishable from that of atmospheric nitrogen treated in the same manner. Confirmation

> of this result would at once prove that the new substance is a manufactured product, which may indeed be present in the atmosphere, but cannot be a new element. With these grave uncertaincies brooding over their new discovery, it is remarkable that Lord Rayleigh and Professor Ramsay should prefer to keep silence, though all doubts might have been settled in almost as many days as months have elapsed since the meeting of the British Association.[12]

The argument that because the new substance was denser than nitrogen it should boil at a sufficiently high temperature to cause the boiling points of chemical and atmospheric nitrogen to differ perceptibly is not sound, and should not have been put forward so confidently, even at that time. In fact, argon boils only 10 K above nitrogen and so the difference in boiling points of 'atmospheric' and 'chemical' nitrogen is only of the order of 0.1 K, a difference that would be unlikely to have been detected in Dewar's experiments on liquefaction. The argument that chemical nitrogen also gave a residual unreactive gas on treatment with hot magnesium was soon found to be false, so removing another prop from the doubters' case.

Rayleigh and Ramsay had two reasons for their continued silence; first, the extreme caution of Rayleigh, the older and more senior man, with more of a reputation at stake, not to publish anything about which there could be any possible doubt, and, second, the fact that they hoped to win the Hodgkins Prize of $10000 offered by the Smithsonian Institution in Washington for the most important discovery relating to the atmosphere that had not yet been published. They had already checked with Samuel Langley, the Secretary of the Smithsonian, who had been present at the Oxford meeting of the British Association, that an oral communication there did not constitute 'publication'. There were 218 applications for the prize but theirs won the day and they received the award from the American Ambassador in London in August 1895.[13] Dewar was to receive the first Gold Medal from the same fund in 1899. He had recently received the Rumford Medal of the Royal Society, 'for research on the properties of materials at low temperatures' on a motion in July 1894, seconded by Rayleigh. It was perhaps significant of the way in which Dewar attracted controversy that when the Council came to decide on the awards on 1 November, all were allocated without opposition except for the Rumford Medal which was awarded only 'after a ballot'.[14]

The report in *The Times* of the meeting at the Chemical Society annoyed Rayleigh who asked Dewar if the last part represented accurately what Dewar had said. Dewar denied that the words were his but he would not write to *The Times* to disown them. Rayleigh's son wrote his father's biography at a time shortly before Dewar's death and he said in the text, published in 1924, that: 'Relations became very strained for a short time, but eventually the matter was allowed to drop by mutual consent, and Rayleigh and Dewar resumed their usual friendly relations when he came back to the Royal Institution in the ensuing February.' The biographer was more frank in his privately annotated copy, for when this version appeared in 1968 it contained the following paragraph:

> I do not quote Dewar's later letters because they are so involved and irrelevant. To argue with him was to put oneself into the same kind of environment as a fly in treacle. He practically avoided the issue altogether by these tactics. J.J. Thomson told me that my father, in his remarks on Dewar at this time, betrayed anger for the first and only time in his (J.J.T.'s) experience.

Thomson was similarly to find himself 'in treacle' over a meeting of the Council of the Royal Society some 20 years later (see Chapter 11).

The physical properties of argon (the name was adopted by those who accepted that the new gas was an element) were soon under study. Rayleigh had originally suggested that he and Ramsay invite Dewar to study its properties in the liquid state, but Ramsay was not happy with this proposal and sent a sample of 300 cc to Olszewski for him to measure its critical temperature and pressure. Ramsay is said to have preferred Olszewski's determination of temperature by the direct use of a gas thermometer rather than Dewar's use of a resistance thermometer as an intermediate device, but his dislike of Dewar may have been the more direct cause for not inviting Dewar to join him and Rayleigh in the project. Ramsay himself undertook the measurement of the speed of sound in the gas at ambient temperature and pressure in order to decide its 'atomicity'. At first it was tacitly assumed that argon had a diatomic molecule, Ar_2, as had the other elemental gases, N_2, O_2, H_2, F_2 etc. If this were so, and since its density as a gas is 19 times that of hydrogen, then its atomic weight is 19 also, and there was no clear place for it in Mendeleev's 'Periodic Table'; fluorine has an atomic weight of 19 and argon could perhaps be squeezed between fluorine and sodium at 23, but the fit was uncomfortable.

The speed of sound in a gas is determined by the ratio of the heat capacities at constant pressure and at constant volume, $\gamma = C_p/C_v$. This ratio depends on the complexity of the molecule; if it is monatomic then γ is 5/3, if it is diatomic then γ is 7/5 or smaller, and so on. The only monatomic gas then known was mercury vapour for which Kundt and Warburg had found a value of 5/3 in 1876. Ramsay measured the ratio of the speed of sound in argon to that in air, for which γ was accepted to be 1.408, that is almost 7/5. For the new gas Ramsay found 1.64 which he took correctly to be 5/3 within his experimental accuracy; so argon is monatomic. This result led to an atomic weight of about 39, which was also a problem for the chemists, since there seemed to no room in the Periodic Table for an element between chlorine at 35.5, potassium at 39.1 and calcium at 40.1.

Rayleigh and Ramsay presented the formal paper on their discoveries to the Royal Society at a meeting on 31 January 1895. This took place in the lecture theatre of London University in order to accommodate the 800 people who wished to attend. This time Ramsay read the paper and it was Dewar who was absent. The presentation was followed by two supplementary papers by Crookes and Olszewski, on the spectrum of argon, and on its critical, boiling and melting points, respectively.[15] The three papers and their discussion were reported at length in *Nature*.[16] The Presidents of both the Physical and Chemical Societies were there and led the discussion. The former, Arthur Rücker, did not share the chemists' worry about the position of argon in the Periodic Table, which he took to be only a convenient classification but one without a sound theoretical basis. Like many physicists and chemists of the time, he had doubts also about the rigour of the argument connecting atomicity with heat capacity, but he accepted the authors' main conclusions with enthusiastic congratulations. Liveing was also among those who also doubted the validity of the argument based on the heat capacity. The President of the Chemical Society, Henry Armstrong, a friend and supporter of Dewar's work, accepted grudgingly the main factual conclusions but condemned part of the paper; it had, he said, 'a portion which was of a wildly speculative nature, viz. the portion dealing with the probable nature of the new element'. Rayleigh was not happy with these arguments and expressed his distress at the behaviour of the chemists. He was later to say that he was glad to get out of chemical discussions and be back among the physicists 'where the second-rate men seemed to know their place so much better'.[17] Dewar and Armstrong soon tacitly dropped their objections in view of the overwhelming evidence for argon as a new element and because of the discovery of helium, another gas of the same kind. Armstrong retired as President of the Chemical Society at the Anniversary Meeting on 27 March 1895 and there, with a good grace and in the presence of Dewar and Ramsay, awarded the Faraday Medal to Rayleigh, who at once acknowledged that Ramsay had had an equal share in the work. Ramsay avoided this point but spoke briefly on his more exciting discovery of terrestrial helium, four days previously.[18]

Helium had been named in 1869 by Norman Lockyer as an element present in the chromosphere of the Sun, and identified by a strong yellow spectral line, very close to the more obvious pair of D lines associated with sodium. The French astronomer Jules Janssen also observed this spectrum and speculated that it was due to a hitherto unknown element. Ramsay had been investigating an inert gas given off by clèveite, a uranium mineral, which he thought at first might be another source of argon, or might even contain a compound of argon. The spectrum did not fit, however, and so the new gas, provisionally named 'crypton', since it was unknown, was passed to Crookes for a more careful examination. Crookes soon realised that Ramsay had found a terrestrial source of Lockyer's helium, and sent a telegram to Ramsay saying 'Crypton is Helium 58749. Come and see it'.[19] Meanwhile other sources of incompletely analysed gases were sought to see what else might be found. One of these was the gas that came up in large quantity with the hot springs at Bath. Dewar, Rayleigh and his son, Robert Strutt, all examined this gas and found it was mainly nitrogen with some carbon dioxide and argon, and a few parts in 10000 of helium.[20] The resourceful Lennox was sent to Bath to recover gas and isolate the helium, the most volatile of the gases.[21] When divers went down into King's Spring in 1968 they found the remains of a wooden structure about eight feet square which had, they believed, been installed by Lennox and his 'Rodebush' assistants.[22] In 1900 the RI paid £159 for Lennox's work at Bath, with smaller sums in each of the next three years.[22] The separation of the helium was not without its hazards; in April 1906 there was an explosion of a mixture of helium and methane in which Heath lost an an eye and Lennox was again injured although not so seriously this time.[23]

The discovery of the remaining inert gases of the atmosphere,[24] neon, krypton and xenon, took place at University College, London and was the work of Ramsay and his colleagues; it is outside our main story. Dewar had, however, claimed that he was also on the track of the obviously 'missing' gas that should, according to the Periodic Table, lie between helium and argon. Ramsay had drawn attention to this gap in his Presidential Address to Section B at the BAAS meting in Toronto in 1897, but he isolated it from air only in June 1898. It is pleasing to note that Travers recalled that Dewar congratulated him warmly at the Chemical Society meeting at which the discovery was announced.[18, 19] Dewar was himself still rejoicing in his recent liquefaction of hydrogen. It is, however, worth noting that this work of Ramsay required liquid air which he did not have at first, and which he could not ask Dewar to provide. It was Hampson who came to his aid and, from May 1898, Hampson brought from Brin's works to University College up to a litre of liquid air at a time. These deliveries were usually in the evening or on a Saturday, when Brin's compressors were not being used for commercially more important duties.

After Dewar's success in liquefying hydrogen in 1898 it was clear that there would be a race to see who could be the first to liquefy helium. The outlook was not promising; its critical point was unknown but the properties of the gas suggested that it lay well below that of hydrogen, 33 K, and also below the normal boiling point of hydrogen at 21 K, or maybe even below its melting point at 14 K, which could be reached by evaporating hydrogen under reduced pressure. If this were so then there seemed to be only two ways of proceeding. The first and simpler would be to cool the helium under a high pressure to as low a temperature as could be reached with hydrogen and then expand it the same way as Cailletet had liquefied oxygen in 1877. If this expansion produced a mist then at least it would be known that liquid helium *could* be prepared, but this route could not, it was thought, produce the liquid in bulk. The second way would be to cool the gas to as low a temperature as was possible and then use Joule-Thomson cooling in an apparatus of the Linde-Hampson type to prepare the liquid in bulk. This would be a formidable engineering problem and it was not known for certain how low a temperature would be needed for the method to work, even in principle. Joule and Thomson had shown that expansion of hydrogen at ambient temperature in their apparatus led to warming, not to cooling, but if the starting temperature is well below ambient then, it was guessed, hydrogen would cool, like all other gases. That this is so was triumphantly confirmed by Dewar's liquefaction of hydrogen in 1898. This problem would, however, be worse with helium. Cooling occurs for each gas only below a temperature now called its Joule-Thomson inversion point, which was not then known for helium. Modern measurements have fixed it at about 52 K at low pressure, but at higher pressures it is at a lower temperature. Such an initial starting point for a Joule-Thomson cooling cycle needs liquid hydrogen as the primary coolant. At first Dewar was optimistic and in 1900 he had convinced himself, probably from the evidence of the isotherms at higher temperatures, that the critical temperature of helium was about 10 K. Dickson was ecstatic, writing: 'I am delighted to hear about the helium. 10° abs will give a B[oiling] P[oint] of about 6⅔°, and perhaps exhaustion another 2° or 3°. The absolute zero will have to look out! It is glorious!'[25] It was not always clear then that no process could reduce the temperature to zero; the 'unattainability of absolute zero' was a consequence of Nernst's 'heat theorem' (or the third law of thermodynamics) which was not established until 1906. When Dewar reviewed the low temperature field in his Presidential Address to the British Association in 1902 he had said 'That the zero will ever be reached by man is extremely improbable'.[26] His reasoning, however, was based more on the ever-increasing practical difficulties with each step towards the zero than to any theoretical restriction of principle.[27]

It was natural that the first attempts at liquefaction would be made with the simpler (and cheaper) Cailletet expansion. This was tried by Olszewski in 1896, before Dewar's liquefaction of hydrogen, and again in 1905. Dewar tried this route in 1901 with helium at 80 atm and cooled by hydrogen at its melting point, 14 K, the lowest temperature then available (Dewar's estimate was 16 K). In the early summer of 1900 he frequently reported seeing mist on expansion, but as the purity of his helium improved the mists vanished; they were probably due to hydrogen or neon; the latter was particularly difficult to remove.[28] When he saw no condensation he deduced that the critical temperature lay below the estimate of 9 K that he had put to Dickson the year before, although he thought that the boiling point was not below 5 K. Kamerlingh Onnes tried the same route at the same time but obtained only doubtful indications of condensation.[29] Travers had no success in 1903.[30]

The conclusion was obvious; there was no short cut; the Joule-Thomson expansion must be used. Dewar started thinking along these lines after his failure in 1901 with a Cailletet expansion. His progress was slow and hindered by a fire at Lennox's Rodebush works in which some of Dewar's equipment was lost, and it was held up for two months, when the electric supply at the RI was changed from 100 volts to 200 volts. These setbacks were followed by his own illness in September.[31] He studied then the adsorption of gases by cooled charcoal which was a necessary step in being able to replace glass vacuum flasks by metal ones. There seems to be no description of the Joule-Thomson apparatus built for helium but it was presumably based on the pattern that had worked for hydrogen. The 'helium regenerator' was in operation by 15 August 1902, according to his laboratory notebook.[32] He wrote to Kamerlingh Onnes in 1903 to say:

> For some time I have been in a very feeble condition of health and have not been able to do [much work?] in the way of research. I have a quantity of helium ready and intend soon to circulate and to do it starting with liquid hydrogen. It is however a complicated and risky business as you well know. I have already lost a cubic foot [28 litres] of helium by the [?] of vacuum vessels during the course of its circulation at liquid air temperatures and I dread any repetition of the disaster.[33]

This loss was in December 1902, for Dickson had written to Dewar in an optimistic vein on 11 December, but sent the next day a letter of commiseration: 'I am awfully sorry to hear about bother with the helium. I hope it can't take as long as you are laying out for it to get the apparatus in working order.'[34] Dewar had been trying an experiment with a 50/50 mixture of helium and hydrogen, presumably because he had found although he could now liquefy hydrogen at will, he was having no success with helium. (It was in such a mixture that the curious phenomenon of 'barotropic inversion' was discovered in 1906. If a gas of relatively high molecular weight is compressed over a liquid of low molecular weight then the liquid floats on top of the gas. Kamerlingh Onnes and Keesom found that this happened with gaseous helium (mol. wt. = 4) and liquid hydrogen (mol. wt. = 2) at 20 K when the pressure is above 30 atm.[35]) In his notebook Dewar recorded that his disastrous loss of helium seems to have come about through a breakage following an obstruction of the high pressure line:

> Now the 50% He and H mixture was passed through the same process the pressure slowly rising to 25 atm for 5 minutes and then to 50 atm and 75 atm. As soon as the highest pressure [was] reached the nozzle and coil began to plug and in a short time was closed up. The solid [that?] had collected [in?] the nozzle began to melt & when the temp. was raised and a [?] liquid formed at the bottom of the V[acuum] Tube when the coil became clear the circulation was repeated when the same stoppage took place as before. The most terrible result was however that *25 litres of gas had been lost in the operations*.[32]

Later he said that 'the whole of the helium that had accumulated during two years was lost owing to the collapse of the glass vacuum vessel containing the regenerator coil'.[36] Had Dewar and Ramsay been on speaking terms it is possible that a combination of Dewar's cryogenic skill and Ramsay's skill in the handling and separation of the inert gases might have led to their collaboration and so to greater success in the London experiments.

Dewar apparently said nothing of the nature of his illnesses in 1901 or in 1903 and it is not clear if they were of the same kind. His minor illnesses are mentioned frequently in his correspondence with Dickson: toothache in 1899, rheumatism in 1900, dyspepsia in 1902, and maybe others that we do not hear about. In 1905 he wrote to Kamerlingh Onnes blaming his illnesses for his lack of progress:

> I had hoped by this time to have settled the helium question [but intermittent?] bad health has been the [great?] obstacle. In my work I have never been able to do anything unless substantially with my own hands and [presumed?] jurisdiction of what is going on. In pioneering work, assistants are a waste. I am ready to start the helium work again, but am interrupted with a hydrogen apparatus that is being tested for the St. Louis Exhibition. If my health breaks down then I would hand the helium [?] to you. This is the best thing I could do for Science.[37]

But when Kamerlingh Onnes asked a year later if Dewar could let him have some of the gas extracted from the Bath springs, he was told that:

> We both want the same material in quantity from the same place at the same time and the supply is not sufficient to meet our great demands. It is a mistake to suppose that the Bath supply is so great. *I have not been able so far to accumulate sufficient for my liquefaction experiments.* If I could make some progress with my own work the time might come *when I could give you a helping hand which would give me great pleasure.* For the present things are in a sad way with me. For the last 4 months I have been seriously ill and quite unable to do any work, so that all research has been [? ? ?].[38]

His ill-health and the disastrous loss of gas in December 1902 did not bring operations entirely to a halt, since he may have had a little gas left and was seeking more from Bath, but much of the impetus had gone from Dewar's research programme. He was a great pessimist, according to Henry Armstrong. Kamerlingh Onnes wrote to him a year later to suggest, rather tactlessly: 'You have told me of your disaster with helium. It has been a fearful one. Could be that you would not have had it by the use of my compressor for pure and costly gases described in Comm. [Phys. Lab. Leiden] no. 54?' Dewar continued with determinations of the density of solid and liquefied gases, some spectroscopy, and improvements in his thermometry, but little advance seems to have been made with the direct attack on the liquefaction of helium. His collaboration with Liveing had come to an end but he had continued with a few of his own his own spectroscopic measurements. He had joined with Moissan in measuring the density of solid fluorine and in studying its reactivity at liquid hydrogen temperatures. He had joined also with Crookes in seeing if extreme cold affected the activity of radium, and with Pierre Curie in the study of the gas evolved from radium bromide – it proved to be helium.[39] A further setback was another serious explosion in April 1906 when Heath lost an eye and Lennox was again injured, although not seriously. But only six pages of his notebook are used to describe the experiments from January 1903 to January 1907, when the helium work was taken up again but with little apparently to show for the effort. All this work came to an end when Kamerlingh Onnes succeeded in liquefying helium in July 1908, just over 10 years after Dewar's success with hydrogen. The end of the work on helium led to a rift between Dewar and Lennox over the difficulties Lennox had had in the purification of helium. The precise occasion for the departure of Lennox is, however, not clear. William Green, who had been a junior assistant in the RI laboratories, said many years later that 'something fundamental snapped in their relationship and work, and the laboratory lost the most outstanding Assistant it had ever known'.[40] Dewar himself expressed surprise at Lennox's departure, and possibly dismay at the behaviour of the only subordinate who seemed to have left on his own terms, and perhaps got the better of him in argument. He wrote on 15 November 1908:

Private and Confidential

Dear Scott,

I have now been [relieved?] of all further trouble about the conduct of Lennox. [?] especially during and after my illness I saw that the engineering interests outside the RI was ruining the scientific work at the RI; and what was done was unevenly perfunctory unless personal interests were involved. It was time therefore to look into matters to defend the credit of [all?] wish the RI. For various reasons I came to the conclusion that Messrs Lennox Co was a bogus affair; not as it was in the [original?] in which other engineers were involved. And thus if this view was correct he was simply passing his own estimates and accounts ... a serious scandal ... I therefore informed in Oct. that he had to elect what his future course was to be, Engineering or Science. To make a long story short the end of it was that he resigned. Now it all comes out because I observe that *he* has sold his business to a Private Co. for the sum of £5000; it being stated that Lennox & Co was simply himself. I blame myself for not discovering this long ago and for being so long suffering and slow to anger. Now that the matter is ended all care must be taken that *no orders are ever again in that locality*. Who the other interested now (and probably formerly interested by having advanced money) are I can't say but in any case a complete change in the matter of that Firm we are under [?] I do not doubt [?] has worked other Firms into the position of indirect orders for Lennox & Co and that the only [?] is *to start afresh with new people*.[41]

Dewar is probably exaggerating his ignorance of the constitution of the company. He must have known that Lennox's partner Reynolds was the brother of the photographer whom he encouraged to roam freely around the RI, and that Reynolds was no engineer but was the administrative side of the partnership. He had been dealing almost every month with the company for 15 years and it is unlikely that he was as ignorant of its circumstances as he said to Scott.

Kamerlingh Onnes's route to the liquefaction of helium was more systematic than Dewar's but even here its final stage was marked by the same kind of error that had affected so many previous claims to have liquefied or solidified the 'permanent' gases. He started with some advantages over Dewar. He had a better knowledge of the isotherms of helium, of van der Waals's law of corresponding states, and a better understanding of the importance of determining the form of the isotherms of the compressed gas in order to estimate its critical and boiling points. He had proceeded more systematically with the training of a body of skilled technicians and glass-blowers, led by Gerrit Flim who played a similar role in Leiden to that of Lennox in London. And, maybe crucially, he obtained his helium from American monazite sands so it was free from neon which was present in the Bath gas used by Dewar. He had successfully liquefied hydrogen by 1906 and was assured of better supplies of both liquid hydrogen (four litres an hour) and of pure gaseous helium than Dewar could command. He built his apparatus in glass and metal, as Lennox also favoured, whereas Dewar chose glass, as he had used for hydrogen, probably because his position at the RI required him to show publicly in the lecture theatre what he had achieved. Kamerlingh Onnes was closing in on the problem of liquefying helium by the spring of 1908. In March he thought that he had a success and sent Dewar a telegram which Dewar passed on to *The Times*: 'Converted helium into solid. Last evaporating parts show considerable vapour pressure, as if liquid state is jumped over.'[42] This spectacular claim was withdrawn the next month; the solid turned out to be hydrogen, as he explained in a long letter to Dewar of 7 April, at the end of which he commiserated with Dewar on his continued bad health.[43] Kamerlingh Onnes had now good grounds for thinking from the measurements of the isotherms at 21 K and 14 K that the critical temperature of helium was about 5 K. Dewar had been hoping for 9 K and if Kamerlingh Onnes was correct then, as he wrote: 'It is clear that any resources that I can at present command would be inadequate to settle the question.' In a letter to Kamerlingh Onnes of 15 April he lamented that the Royal Institution had no money for such expensive work, but at least his health was improving and the 'Medical Faculty' had sent him to the seaside to recuperate.[44] By 1908, however, Dewar's attempt to liquefy helium was at an end. Kamerlingh Onnes finally achieved this success on 10 July 1908, but again there were initial doubts, and Dewar was asked to withhold the news until further tests were made. These were completed by the morning of 15 July and it was on 17 July that Dewar was able to confirm to the British public that helium had been liquefied; the last element had been reduced to a condensed state.[45] Dewar sent his hearty congratulations. In the immediate aftermath of his success Kamerlingh Onnes himself was taken ill but he was able to submit a report, read by Dewar, to the British Association meeting in September in Dublin.[46] He had recovered a few weeks later and was sufficiently well to go to the Premier Congrès International du Froid which met in Paris on 5 October. Dewar declined to attend and the British delegation was led by Ramsay. This meeting led to the setting up in January 1909 of a more permanent organisation, the Association Internationale du Froid; Dewar was to serve on its scientific committee. The first Congress had resolved to assist the Leiden laboratory in inviting outside workers to make use of its facilities – a move that may have led to some delay in Kamerlingh Onnes's own work. The second Congress in Vienna provided a grant to support this work and recognised the unique place that the Leiden laboratory occupied in the cryogenic field.[46]

After the liquefaction of helium there was still a mystery about the existence of solid helium since Kamerlingh Onnes had found that the substance remained liquid down to a vapour pressure of 10 mmHg. He was, however, looking in the wrong place; helium remains a liquid down to the absolute zero unless the pressure is above 25 atm, when it solidifies, but this unusual phase behaviour was discovered by Keesom only in 1926, after the death of Kamerlingh Onnes.

Liquid helium was first produced outside the Netherlands in 1923 by John McLennan at Toronto, using equipment built on plans supplied from Leiden. It was first seen in Britain in 1932 when Gerrit Flim brought two flasks of liquid helium from Leiden to the RI for a Friday Evening Discourse by McLennan. The vacuum flasks containing the helium were surrounded by ones containing liquid air which, in turn, were placed in metallic vacuum flasks that also contained liquid air. These were flown from Leiden in a two-seater aircraft. An electric currrent, started at 3 p.m. in Leiden in a superconducting lead coil immersed in the liquid helium, was still flowing at 9 p.m. the same evening at the RI. In this way both liquid helium and the phenomenon of superconductiviy were first demonstrated in Britain. The next year liquid helium was produced in Britain for the first time by Kurt Mendelssohn in the Clarendon Laboratory at Oxford.[47] The apparatus he built followed the pattern used by Franz Simon in Breslau (now Wrocław in Poland) which made use of the Cailletet expansion but in a way that preserved the fluid that remained behind the valve and allowed the liquid part to be collected.

The enmity, for no lesser word will do, that arose between Dewar and Ramsay from the discovery of argon and helium and the liquefaction of hydrogen, was probably of long generation. They met first in September 1876 at the meeting in Glasgow of the British Association, when both were young organic chemists. Dewar was then 33 and Ramsay 10 years his junior. Six months later they were both candidates for the Fullerian Professorship at the RI, when the Managers, not surprisingly, chose Dewar. After Ramsay had moved from Glasgow to Bristol and Dewar from Edinburgh to Cambridge and London, both changed their fields of research to the physical study of gases and liquids. In 1886 Dewar was asked by the Royal Society to referee a paper by Ramsay (not yet a Fellow) and his colleague Sydney Young, on the pressure-volume-temperature relations of ethyl ether. Dewar thought that their work was too similar to their previously published papers, and to the work of Thomas Andrews. His own assistant, Ansdell, had made similar measurements but these had appeared only in brief in the Society's *Proceedings*; he saw nothing in Ramsay and Young's work that justified publication in the more prestigious *Philosophical Transactions*. The manuscript and Dewar's report were then sent by Stokes for a second opinion to Crum Brown who recognised the value and quality of the work, and who recommended its publication. There is no evidence that Ramsay knew that Dewar was the first referee, but he may well have guessed.[48] Open warfare broke out at the Chemical Society in December 1895, when Ramsay tried to cap Dewar's success with his new apparatus for liquefying air and, he believed, hydrogen also, by saying that he, Ramsay, had heard that Olszewski had obtained not only a spray of liquid hydrogen but also the liquid in bulk.[49] No claim from Olszewski was forthcoming, but when Dewar announced to the Royal Society his own successful liquefaction of hydrogen in bulk in May 1898, Ramsay repeated his claim for Olszewski's priority. Dewar challenged him to produce evidence for Olszewski's success, which Ramsay could not do, and was forced later to tell the Society that Olszewski had now written to him to say that he had never made, or claimed to have made, liquid hydrogen in bulk. Ramsay never put his retraction into print and Dewar gave his own account of the matter in a Friday Evening Discourse at the RI.[50] Ramsay was clearly in the wrong in this affair, but he in turn had been justified in sharing Rayleigh's annoyance over Dewar's behaviour over the identity of argon in 1894. It must have been a rather strained eight-course dinner when Dewar and Ramsay sat opposite each other in the Hotel Métropole in 1896 at the celebration of the Jubilee of the Chemical Society.[51]

The antagonism spread into the public domain early in 1897. On 18 February the Council of the Chemical Society proposed that Dewar be the next President, to succeed Vernon Harcourt. Ramsay was proposed as a Vice-President and was awarded the Longstaff Medal for his discovery of helium and for his share in the discovery of argon. The choice of Dewar was not received with the usual formal approval of the Society and by the next meeting, a month later, about 112 Members had proposed Ramsay in his stead. The dissatisfaction with Dewar's nomination arose in part from the reputation he had acquired as an argumentative man, but also from the feeling that the Council was, in effect, a self-perpetuating group that took no notice of the views of some of the younger members. Ramsay had probably also acquired a better reputation at the Society by his conscientious attendance at its meetings and his participation in the discussions. Dewar usually came only when he was directly concerned with the topics to be presented. The opponents for his election to the post included substantial contingents from Cambridge, Oxford and, inevitably, from University College, London. Their spokesman was often M.M. Pattison Muir of Caius College, Cambridge, who had led the support for Ruhemann in his row with Dewar some seven years earlier. The case for Dewar was led by his friend Henry Armstrong who opened the defence on 23 March with a letter to *Chemical News* saying that Muir was not a man whose opinion was to be trusted, as was clear from his letters to *Nature* two years previously in which he had intervened in the row between Dewar and Olszewski. Norman Collie, an Assistant Professor at University College, said that Ramsay had not known of his impending nomination, but Armstrong did not believe a word of it. He had to contend also with criticism of Dewar's alleged misconduct in the case of 'the celebrated Edison Swan patent action' (see Chapter 7) but here Dewar had been in the right. The election was held at the Anniversary Meeting on 31 March and Dewar was chosen. No figures were released, neither then nor at the next regular meeting of the Society on 1 April, when Dewar was in the chair. The scrutators were unsure of the validity of some of the voting papers, although not about the outcome of the election. Dewar was against publishing the figures but the Council called for a recount to satisfy the natural wish of the members. They were eventually released on 6 May: Dewar 166, and Ramsay 152. This battle between the two Scots attracted Ian McCrockett (clearly a pseudonym) to send an account of the fight to the *Chemical News* in Scots dialect, entitled 'Who shall be hen-wife?' It concluded: 'There was a confused struggle. Wullie squealed and kicked, the Chem. Soc. awoke with a start, and found Jemmie installed as hen-wife, and the kailyard resuming its normal gentlemanly demeanour.'[52] Dewar took his duties seriously and chaired every meeting during the two years of his presidency. Indexes to the journal were prepared; a new editor was appointed and the journal was issued on time and not several months late as had become the custom. At the end of the day the Society had made a good choice. He was succeeded in 1899 by Thomas Edward Thorpe; Ramsay had to wait until 1908 for his turn.

The discovery by Ramsay, Collie and Travers of the other rare gases, neon, krypton and xenon, was the occasion of unpleasant anonymous letters criticising Ramsay's work and alleging that he did not give proper credit to Rayleigh. The author has never been definitely identified; one of his letters is marked as coming from London but that still leaves a wide field. Then and now, however, the consensus was that Armstrong was the author. The writer used the pseudonym 'Suum Cuique' and his letters appeared in *Chemical News*, starting on 17 June 1898.[53] His main scientific point was that Ramsay was too ready to claim the discovery of new elements on the evidence of spectral lines whose identity was uncertain. He wrote, for example: 'If spectra of mixtures of unknown composition are to be accepted as discovery of new elements then Liveing and Dewar discovered krypton four years ago and Dewar discovered a new gas last November to which he has not yet given a name.'[54] In disentangling the mixtures of the five inert gases Ramsay blundered at one point when he claimed that there was a partner to argon, with a similar atomic weight, which he christened 'metargon'. For some years he had believed that the new group of the rare gases contained closely related triplets, like those of the metals: iron, cobalt and nickel, etc. Dewar correctly doubted the existence of metargon.[55]

Chapter 10

The Davy Faraday Research Laboratory[1]

In 1843 the Managers had considered a proposal from Brande and Faraday to extend the remit of the RI to include a school of practical chemistry. There was, however, not enough space in 21 Albemarle Street for the plans to be carried out and the proposal was dropped. A solution came into sight in 1892 when Ludwig Mond bought the freehold of 20 Albemarle Street. By then there were other institutions in London that were offering training in chemistry to prospective students but, Mond thought, there was still a need for laboratory space for trained independent chemists to carry out research. Professors generally had one or more assistants but there were no research schools where a man could aim for a higher degree in the way that was then possible in Germany, and still less, for independent or semi-independent research workers of the kind that we now recognise as post-doctoral fellows. It was this need that Mond, now a wealthy man and living in London, aimed to meet. His formal association with the governance of the RI began in 1889 when he joined the Board of Visitors, and was strengthened in 1892 when he became a Manager for the next two years. He was re-elected[2] in April 1894 and almost at once became a Vice-President and a member of all Committees except for the Library Committee.[3] Meanwhile, in April 1893, Frederick Abel had applied to the Managers for 'permission to be allowed to make use of extracts from a Report on a proposed School of Practical Chemistry at the Institution, prepared by Professors Brande and Faraday in 1843'.[4] Clearly the ground had been well prepared for the greatest single benefaction that the RI had received. On 2 July 1894 Mond laid before each of the Managers an offer which he explained in the following terms:

> It has been my desire for many years to found a Public Laboratory which is to give to the devotees of pure science … the facilities necessary for research in Chemistry and more particularly in that branch of the science called Physical Chemistry. … I have come to the same conclusion as the Promoters of the scheme of 1843, viz.: that such a Laboratory would still have the greatest prospect of success under the aegis of the Royal Institution.[5]

The offer comprised the transfer of 20 Albemarle Street to the RI, payment of the cost of the alterations needed for it to be incorporated into the existing laboratories at No. 21, £500 p.a. for running costs, and £1200 p.a. for salaries and other costs. The last sum was conditional on the 'productivity' of the research workers being maintained. The facilities were to be offered to trained chemists who must provide for their own salaries and research expenses beyond the basic provision of space, heating, light and water. The administration of the Laboratory was to be in the hands of a Committee appointed by the Managers, and under the day-to-day authority of the Resident Professor (that is, Dewar) assisted by a Superintendent. Ludwig Mond's son Robert was to be the Secretary of the Committee. All this was agreed with the Managers. Over the summer of 1894 Dewar and Wiliam Flockhart, the RI architect, drew up plans for the work to be carried out. The extra space allowed not only the creation of the Davy Faraday Laboratory, as it was to be called, but also for the expansion of the existing Chemical Laboratory, needed by the scale of the apparatus used in Dewar's work on the liquefaction of hydrogen. The area at the back of No. 20 was enclosed to become the 'New Laboratory' for Dewar's use but later became the workshop. The attachment of the new building allowed for the extension of the Library on the first floor and of the Dewars' quarters on the second floor. Dewar himself contributed £250 to the last item; the total cost to Mond was about £45000 for the building works and laboratory equipment, and £65000 for an endowment.[1] The plans were complete by March 1895 and the works were finished in the autumn of 1896, when Mond had retired as a Manager.[6]

The Committee was to be Rayleigh, Dewar and Bramwell, as officers of the RI, with Abel, Abney, Frankland, Ludwig Mond (replaced by Crookes in May 1895) and Müller (who seems to have declined appointment), with Robert Mond as Secretary and Henry Young, the Librarian of the RI, as Assistant Secretary.[7] He was already the Assistant Secretary of the RI and was paid an extra £50 (raised to £100 in 1899) for taking the additional role at the Davy Faraday Research Laboratory. Before the Laboratory was opened Dewar had been joined by Rayleigh as a Director, although the administration remained substantially in Dewar's hands. Neither he nor Rayleigh were formally paid for these responsibilities,[8] but Dewar's 'additional' salary, that is, the part not paid from the Fullerian endowment, was raised from £300 to £650 from 1 January 1896, and Rayleigh received £300. The day-to-day running of the laboratory was put into the hands of a Superintendent or Assistant Director – the use of both titles was to lead to trouble later. Two candidates were considered, one of whom may have been John Shields,[9] who worked for Mond, but, after interview by Rayleigh and Dewar, the other was chosen. He was Alexander Scott, Dewar's Jacksonian Demonstrator at Cambridge,[10] who took up his new duties on 31 July 1896 and began the task of appointing junior staff when the building works were completed in the autumn. The formal opening by the Prince of Wales took place on 22 December, an event that was recorded with an article in the *Illustrated London News*.[11] Dewar gave a lecture of which the illustration in this article is entitled 'All we have to fear is an explosion'; an apt choice of comment in view of Dewar's record of explosions and injuries to himself and his staff.

The new laboratory occupied an unusual legal position; it was formally separate from the RI itself with its own trust deed, but it was physically embedded in the now combined building of 20 and 21 Albemarle Street, and its administration was in the hands of the two RI professors, Dewar and Rayleigh. How far it was under of ultimate control of the Managers of the RI was a matter that was left vague and had eventually to be resolved by taking counsel's opinion. Another future source of confusion was the degree to which the junior staff of the DFRL were expected to work also in the RI itself. The decision of who should be admitted to conduct research there was nominally in the hands of the Committee, but, in practice was in the hands of the two Directors. There were six appplicants for admission on the opening day of 1 January 1897, of whom two were accepted. One was Joseph Petavel who was already well-known to Dewar in his role of assistant to Ambrose Fleming, Dewar's collaborator at University College. The other was Meyer Wildermann about whom little seems to be known. He had a Ph.D. from Freiburg and, unusually, a B.Sc. from Oxford, a junior research degree that had been instituted only two years earlier. He was to work in the DFRL altogether for 12 years.

From the opening of the Laboratory in 1897 until Dewar's death in 1923 there were 123 applications, of whom 65 were offered posts and 58 took up the offers. Of the 58 who worked there 15 were or became Fellows of the Royal Society. Scott was elected to the Society in 1898, soon after his appointment as Superintendent. Perhaps surprisingly Dewar was not among those who signed his certificate in 1894. The workers' contributions to science lay mainly in chemistry; their work was important in its day but little of their research led on to major achievements after their terms in the Laboratory had come to an end. Dewar, Rayleigh (who worked mainly at Terling) and Scott were essentially lone workers and interactions between the workers themselves was apparently minimal so there was no nucleus around which any concerted programme could form. This outcome was, however, presumably what Mond envisioned when he had set up what William Bragg's daughter later called 'temporary hotel accommodation' for independent research workers.[12] The trust deed did not encourage any attempt at combining the efforts of the two sister departments of the RI. Things were to change when William Bragg became Fullerian Professor of Chemistry, Superintendent of the House, and Director of the Davy Faraday Research Laboratory, on Dewar's death in 1923. He was to preside over a well-knit research school that took both laboratories into the new field of the determination of the atomic structure of crystals by X-ray diffraction.

Long before Bragg took over, the reputation of both the DFRL and the RI itself had fallen into decline. In the case of the former the trouble started with the falling-out of Dewar and Scott. The first dispute arose indirectly from a row between Petavel and Dewar in October 1900, over what were called some 'very objectionable charges' that were levelled against Dewar by Petavel who had been given some minor administrative duties in the DFRL as Assistant Director of the Physical Laboratory.[13] The nature of the charges was not revealed but Dewar naturally took offence and was backed up by both Robert Mond and Rayleigh. Petavel was excluded from the Laboratory until he apologised and Scott became involved when it became clear that he knew of Petavel's offence but had neglected to tell Dewar that he had known. He wrote, 'I admit unreservedly that I see clearly that I ought to have let you know that I was aware of the strained relations existing between yourself and Mr. Petavel'. Three years later, in December 1903, Petavel reclaimed his position after apologising for 'the want of tact on my part' and after he had been chosen by Dewar to take the liquid-hydrogen apparatus to the St Louis Exhibition in the summer of 1904, and to deliver some 80 expository lectures that were given there.[14] A more serious dispute between Dewar and Scott arose in 1910 in a way that recalled the dispute between Dewar and Ruhemann at Cambridge some 20 years earlier. The trust deed of the DFRL made it clear that the work there was to be disinterested science and that commercial work was not permitted. It seems, nevertheless, that Dewar was allowing some of the facilities to be used to support his extensive commercial and legal work, either by work on the premises or by 'borrowing' DFRL staff to do such work within the RI. Scott took his duties as Superintendent seriously and objected to such practices. It was, however, ostensibly on grounds of the expense of running the DFRL that Scott was removed from the post of Superintendent. At the meeting of the Committee of 7 November 1910 it was resolved 'That the Office of Chief Assistant or Superintendent of the Laboratory be terminated at the end of the Lent Term, and that no successor be appointed'.[15] Such a resolution amounted to giving Scott four months' notice of his dismissal. A complaint to Dewar, who had been backed by Robert Mond, led nowhere and so, two weeks later, Scott objected in a letter to the Managers:

> This dismissal, for such it amounts to, fills me with astonishment, not only because of its suddenness and of the absence of any communication with me, but also because it is totally at variance with the nature of my appointment as explained to me before I accepted it, both by the late Dr. Ludwig Mond, and also by Sir James Dewar, who both explained that the appointment was for life, and that what was wanted was a man ready to devote his life to research. … But for these assurances I certainly should not have given up my appointment at Cambridge, and I doubt if this can have been explained to you by Sir James Dewar, who knows this quite well, although I gather that he denies it or has forgotten it now.[16]

He quoted from a letter he had received on 5 July 1896, at the time of his appointment at the DFRL, in which Dewar explained that: 'The general tenure would be like the Cambridge one. Although the Professors are appointed annually,[17] yet the assistants simply go on.' Scott thought that the alleged financial crisis in the affairs was merely an excuse, and later wrote to the Managers to say:

> This plea is so weak that it will not bear examination for a moment, and I am therefore driven to seek some other reason for the attack that has been made upon me. I can suggest nothing but that I have given offence in objecting, and as far as possible in enforcing, that the use of the Laboratory, its materials and assistants should be solely to pure research work, for which the Laboratory was established, and should not be employed for patent or legal work for the enrichment of individual persons.[18]

The members of the DFRL Committee and the Managers of the RI can have been in no doubt that Dewar was the 'individual person' whom Scott had in mind. He had, however, misjudged the character of his adversary for Dewar replied at once with a line of argument whose force could not be denied:

Dear Dr Scott,

Instead of carrying out my request of preparing a specific Lab^{y} Return you have thought after legal consultation [to send?] a reply copy of a Document you addressed to the Managers of the RI on 1st Feby along with a covering letter explanatory of the meaning of a particular paragraph in the said Document; which you say is the construction every Manager would put upon it. ... The question remains who is the personality to whom reference is made when you say 'I have given offence in objecting and as far as possible in enforcing that the use of the Laboratory and its materials and assistants should be confined solely to pure research work for which the Lab^{y} was established and should not be employed for Patent or Legal work undertaken for the enrichment of individual persons.' Clearly the person in question is outside the class of Workers otherwise no reservation about 'as far as possible' would have been intruded; and further he must have direction and control of other *asistants in the Laboratory*. Who can this be if the Director is excluded? Let me have a reply to this letter along with the return I have requested in my letter of 11th March.

Yours faithfully, James Dewar

PS A couple of days is ample to do what is required.

JD

Dewar had called his bluff, for Scott could only reply:

Dear Sir James,

The last paragraph but one in my letter of February 1st addressed to the Managers, did not refer to you, and I therefore decline to discuss this subject further. The Senior Laboratory Assistant to whom you have given directly all instructions regarding any work done for you informs me that as far as he knows there are no records of any work done for you in any of the Laboratory note books, but that all notes concerning such work were delivered to you at once,

Yours faithfully, Alexander Scott[19]

Had Dewar taken Scott to court for what might be a libel in Scott's first letter it is unlikely that a judge or jury would have accepted Scott's claim that Dewar was not the man meant; Dewar's argument on this point seems unanswerable. Dewar did not, however, follow this route but contented himself with a formal refutation: 'The assertion by you that any Legal or Patent work done in the Lab^y was done "for the enrichment of individual persons and in no way was contributory to the benefit of the corporate interests", aggravates the offence.' Throughout his career Dewar had been careful to keep all the papers for his legal and consulting work out of the records of the RI and perhaps he feared that if he did sue Scott he might be forced to produce in court papers that might have supported Scott's charges.

Counsel's opinion was taken on the validity of Scott's claim that he had tenure in his post, and Sir Robert Burley held that the Managers had the right to end his appointment but that he was entitled to reasonable notice, for which he suggested the payment of a year's salary. The problem also raised the question of the degree of authority that the Managers of the RI had over the conduct of the DFRL in view of the separate trust deed that governed the affairs of the latter; again Counsel advised that the Managers had the ultimate right to determine the way the DFRL was run. Clearly either Scott or Dewar had to go, since they could not now work together, and the Managers, wherever their own sympathies might lie, probably felt as a body that Dewar was the star of the RI and could not be lost. Rayleigh's position was not known. He had resigned his Professorship at the RI in 1905, after the death in December 1904 of his valued research assistant, George Gordon, but he continued as a Director of the DFRL until his own death in 1919. Perhaps he thought the matter was merely another row between two argumentative Scots, similar to that between Dewar and Ramsay, but he seems to have done nothing to help resolve the issue, although he had offered to help earlier in the dispute with Petavel.[13] He had made it his custom to invite scientists and their wives for weekends at Terling Place to see his laboratory there. Scott came in 1901 but the Dewars were not asked until 1904 when Rayleigh was about to resign his RI chair.[20] The timing of the two visits may be some indication of where Rayleigh's sympathies lay. Dewar had hoped that Rutherford might succeed Rayleigh but he declined, and the new Profesor of Natural Philosophy was J.J. Thomson who also took no part in trying to sort out the problems but simply sent later a postcard to Scott to say that he was glad that the affair had not disturbed the good relations between them.[21]

Scott left his post in April 1911 but the legal arguments went on for several years over compensation for his loss, over his right to continue in the DFRL as a regular worker, and over the ownership of his apparatus and materials.[22] The case was concluded only after the start of the First World War, when Scott continued with his research privately until he was employed by the British Museum in 1919. During his period of unemployment he was much involved with the affairs of the Chemical Society, being first a Secretary, then the Treasurer from 1904 to 1915, and finally the President from 1915 to 1917. The external scientific world naturally sided with Scott rather than Dewar. Tilden wrote to Scott to say: 'I have just been reading the Statement of your case which you are sending to the Members of the RI. I feel sure that you will have the cordial sympathy of the whole of the scientific world.' Ramsay was naturally one of those who took Scott's side; he wrote:

> I have read your statement, & all I can say is that I wonder how a set of men, each of whom in his individual capacity is an honourable man, can have jointly constituted themselves a gang of thieves. Dewar has a most inexplicable influence on others; he is a clever man. I was sitting next [to] Ballantyne, who used to devil for him, at dinner not long ago & said that Dewar had some of the virtues, hospitality, kind impulses etc.; to which Balantyne replied with heat:- 'You are entirely wrong; he is an incarnate devil.' And me thinking to say something nice about a friend!
>
> You must certainly recover your preparations, and your mercury. It appears to me that Robert Mond has got tired of spending money on the D.F. & has wished to end it. But he might have gone straight to work about it and told you so. … [He made the point that Scott was under the Directors of the DFRL and not of the RI itself and then added:] I have full right to dismiss an assistant at sight; and it might be argued that that was your position.[23]

Scott's departure was followed by the resignation of two of the assistants in the DFRL, leaving Dewar, helped by his own assistant, William Green, to run the place himself.

Late in the day the Board of Visitors became involved at the instigation of its former member, John William Gordon, a barrister with an appetite for cantankerous disputation that matched Dewar's. He had become a member of the RI in 1899 and was on the Board of Visitors from 1909 to 1911. In October 1914 he sent a memorandum to the President, the 7th Duke of Northumberland, in which he complained that Dewar and Robert Mond had dismissed Scott 'for no cause assigned' and that this action had cost the RI £1000. The Visitors had disapproved and had tried to add a passage to their Report for 1912 to express their opposition. The President had refused to call a Special General Meeting to discuss the matter and so no Report for that year was approved. Nevertheless a Report was issued purporting to have been signed by the Visitors who had not voted for it. So Gordon then asked that Dewar should report on the work of the Laboratory, but this Dewar refused to do. Gordon had earlier in the year had asked that the Professor of Chemistry should report to the Anniversary Meeting of the RI itself on 'all discoveries and improvements in Chemical Science made in the laboratories in the preceding year', but the Managers had ignored this request 'since nothing similar had been done for many years'.[24]

Chapter 11
Decline

In 1902 Dewar turned 60 and was at the height of his career. The liquefaction of hydrogen had established his claim to be the leading authority on all matters cryogenic, and he had every hope of soon being able to liquefy helium also. He was fortunate in his colleagues. His fellow professor and co-director of the Davy Faraday Laboratory was Lord Rayleigh, perhaps the most productive and wide-ranging living physicist, their immediate assistant was Alexander Scott, himself an excellent chemist who ran the Davy Faraday Laboratory most efficiently, and Dewar's leading technical assistant was Robert Lennox, who probably knew more about the practice of gas liquefaction than any other British chemist or engineer. The Secretary and Treasurer of the RI were William Crookes and James Crighton Browne, who both supported him, at least for the time being. Two years later Dewar's contributions to British science were publicly recognised by the award of a knighthood. He had, he said, previously declined this honour but now that Ramsay had been knighted in 1902 he must have changed his mind. Only his health might have been causing him to worry.

That year, 1902, he was elected President of the British Association which was to hold its meeting in September in Belfast. He travelled there by way of Kincardine where he and his wife stayed at the Unicorn Inn. In Belfast he treated the audience to a massive Presidential Address[1] in which he first followed custom by reviewing the city's place in the history of science and then by paying tribute to Tyndall, who had presided when the Association had last met there in 1874. He repeated the oft-repeated complaint of how British scientific education was falling behind that on the Continent, a theme that had been common to many presidential addresses, at least since Frankland had been president of the Chemical Section in 1868, the first year that Dewar had attended a meeting of the BAAS. Dewar discussed at length Tyndall's controversial views in his famous Address in 1874 in a way that is now of more interest for what it shows of Dewar's views than of Tyndall's. He commended Tyndall's much quoted declaration that 'We claim and we shall wrest from theology the entire domain of cosmological theory'. He commended also Tyndall's robust defence of Darwin's theory of evolution, but was careful to acknowledge Tyndall's (and so his presumably his own) view of the limitation of the proper claims of science: 'the whole process of evolution is the manifestation of a power absolutely inscrutable to the intellect of man'. He was then unfortunate in his assessment of the immediate future of scientific theory. He regretted that the present day had nothing as dramatic to show as the theory of evolution. He wrote: 'It was Tyndall's good fortune to appear before you at a moment when a fruitful and comprehensive idea was vivifying the whole domain of scientific thought. At the present time no such broad generalization presents itself for discussion.' Dewar was, however, not the man to perceive the first glimmerings of quantum theory, and indeed there were many men better informed than he was about theoretical physics who did not recognise what was underfoot at this early date. Planck's theory of radiation was developed at the close of the nineteenth century as an attempt to solve the problem of the spectrum of the distribution of black-body radiation in a system at thermodynamic equilibrium. The development of this theory required him to treat the energy of radiation as 'quantised'. As he wrote in his paper of 1900: 'But I regard *E* [the energy] – and this is the essential point of the whole calculation – as made up of a completely determinate number of finite equal parts.'[2] Rayleigh knew of this paper but neither he nor Planck appreciated then what a revolutionary theory it was to inspire. It seems that Dewar did not discuss theoretical physics with Rayleigh but relied instead for such advice as he thought he needed on his brother-in-law, Hamilton Dickson, whose interests were confined to the conventional physics of the nineteenth century. No sooner, however, had Dewar delivered his pronouncement on the absence of a 'vivifying' theme in present-day science than Joseph Larmor read the next week a paper to the Mathematics and Physical Division of the meeting 'On the application of the method of entropy to the problem of radiant energy' in which he described Planck's new work. This paper appeared only in abstract in the *Reports* but in the same year Larmor published in the 10th edition of *Encyclopaedia Britannica* an article on radiation in which he dealt again with Planck's theory.[3] No blame can be attached to Dewar's verdict on the lack of a 'vivifying' idea in current physics or chemistry when the great theoretical authorities of the day, Planck, Rayleigh, Larmor, Lorentz and others, had not yet fully woken up to the consequences of the bomb being laid under the structure of classical physics. It is only we, with possibly smug hindsight, who can see how mistaken was Dewar's prediction. Kelvin never engaged with such new ideas but he was aware that there something wrong with current understanding of the science of energy, an unease that he had expressed in a Friday Evening Discourse in April 1900.[4] Both he and Dewar were more receptive to the novel and better-known experimental discoveries that accompanied these theoretical explorations. The 1890s had seen the discovery of the electron, x-rays, radioactivity and the group of inert gases, and Kelvin retained an interest if not always an understanding of the work behind these discoveries. In June 1903 he was staying with Crum Brown in Edinburgh when he wrote to Dewar: 'Have you done anything for radium at very low temperatures? Does it cease to glow visibly when greatly cooled, and begin glowing again when warmed towards or beyond 0 °C? If it behaves so we could see (I believe) a dynamical explanation of its radiation by the shaking out of electrions [*sic*] from its atoms impulsively, and their creeping back towards the normal quantum.' A few weeks later, after he had discussed the matter with Pierre Curie on his way through Paris, he wrote: 'Meantime I suppose that we may consider as certain that a gramme of bromide of radium emits 90 gramme-calories per hour. … I cannot think how Rutherford and others can complacently attribute the emission of heat by radium to energy stored in the substance. … We are all at our wits' end in respect to radium.' Kelvin saw here a threat to the first law of thermodynamics! Curie found that the emission of heat was, if anything, stronger at low temperatures. Dewar and Crookes had tackled together the problem of the emission of radiation by radium and found again that there was no appreciable difference between the quantity emitted at low temperatures from that at room temperature.[5] But when Dewar came, in the main part of his British Association Address, to the customary review of recent work he deliberately restricted himself to his own field of cryogenics. Here he eschewed any theoretical or thermodynamic discussion of the field except to remark that van der Waals's law of corresponding states was the most important advance in thermodynamics since Carnot's time, a comment that ignores the work of Clausius, Kelvin and Gibbs! Even in this Address Dewar valued the law not for its molecular implications, as had been drawn out by Kamerlingh Onnes,[6] but for its practical value in estimating the physical conditions under which one might hope to be able to liquefy a gas. Kelvin wrote to him to congratulate him on a fine address, but to chide him gently for confusing Kelvin's first attempt at defining an absolute scale of temperature in 1848 with his later, and now accepted, version of 1854.[7]

Dewar's position as leader of the cryogenic field led to an invitation to write the long article on liquid gases for the 10th edition of *Encyclopaedia Britanicca* in 1902. His article was far from impartial, with his own work being described at length and supported by two photographs and an engraving. Hampson received but a single passing mention and no discussion. When he revised the article for the 11th edition in 1911 then the photographs and engraving were dropped but even Kamerlingh Onnes's liquefaction of helium in 1908 was made to appear as a consequence of Dewar's showing the way in his 1902 article.

Other honours, formal and informal, flowed his way around the turn of the century; he was chosen to represent the RI at the coronations of Edward VII, who as Prince of Wales had been a keen supporter, and of George V.[8] His position at the junction of physics and chemistry was reflected in his receipt of the Rumford Medal of the Royal Society for work on 'the thermal and optical properties of matter' in 1894, and of the Davy Medal for 'a recent discovery in any branch of chemistry' in 1909. This position was reflected again in his nominations for a Nobel Prize.[9] From 1903 to 1913 he was nominated nine times, five times for physics and four for chemistry. The first nomination was in chemistry in 1903 by the amateur achaeologist John Lubbock (Lord Avebury from 1900) but this was disallowed since the nomination was not sufficiently specific in describing the work put forward. Avebury tried again in 1909 but again his proposal was disallowed on the same grounds; the Chemistry Committee was more insistent on this point than the Physicists. The other two Chemistry nominations came in 1910 and 1911 from Vincenz Czerny, a surgeon in Heidelberg. He was catholic in the range of his nominations; they ran from Boltzmann to the Wright brothers and von Zeppelin; it is not clear why he chose Dewar. The five physics nominations were more orthodox. In 1904 the Russian theoretical physicist Nikolai Umov nominated both him and Olszewski, and in 1905 the astrophysicist Orest Khvol'son nominated Dewar. There was then a pause until 1912 when the physicist, Emil Warburg, submitted a proposal in which he suggested that the prize be shared between Dewar and Kamerlingh Onnes. Such a proposal was repeated the next year by Philipp Lenard, the Laureate of 1905, when Warburg also tried a triple nomination of Linde, Dewar and Kamerlingh Onnes. The last of these was successful as the solo Laureate for 1913; it seems that liquid helium in 1908, backed by superconductivity in 1911, had trumped liquid hydrogen in 1898. The most striking aspect of these nominations is the absence of any support for Dewar from the British and Dutch physical and chemical communities. During Dewar's lifetime there were nominations for the Physics prize from Rayleigh, Kelvin, J.J. Thomson, Kamerlingh Onnes and van der Waals, and from Ramsay for Chemistry, but none was in favour of Dewar.

At the start of the twentieth century Dewar was well-situated at the heart of the British scientific 'establishment'. He spent most of the year in London where he was comfortably housed at the expense of the RI, and he had his own house in Cambridge which he used when he came down to give his statutory lectures. He was a married man with no children but two salaries, either of which would have supported him and his wife. He also earned substantial fees from his commercial and legal work and so was able to live the life of well-off London gentlemen. He and his wife were noted for their generous hospitality, as is clear from the letters of their guests. Their flat was heavily furnished in the custom of the time with expensive pictures and tapestries. He took a great interest in the collection and display of portraits and sculptures of earlier inhabitants and officers of the RI and these were displayed in his flat and throughout the domestic and public spaces of the building.[10] Ludwig Mond was again the principal donor. The best-known and certainly the largest painting was acquired with his support in 1906, it is by Henry Jamyn Brooks and purports to show Dewar lecturing on the liquefaction of hydrogen in 1904, but in fact the painting is not based on an actual occasion and the audience is a made-up collection of some major and some minor figures in contemporary science, arts and public life.[10]

He had maintained his early interest in music and still played his fiddle, accompanied on the piano by his wife, after the day's laboratory work was over. He had added to his music a taste for the theatre and the opera. In Albemarle Street he was just around the corner from 15a Grafton Street where Henry Irving lived for some years. He and Irving were friends and part of a circle that included Bram Stoker, who managed Irving's Lyceum Theatre, Ellen Terry his companion and leading-lady, and Alexander Mackenzie who became the Principal of the Royal Academy of Music, and who was also a composer who wrote music for Irving's productions. Dewar served for 34 years on the Board of the Royal Academy of Music; he was a member of the Garrick Club and the Athenaeum to which he had been elected in 1884 under the 'fast-track' rule for distinguished candidates, on the proposal of George Liveing and Warren de la Rue. He generously supported several struggling young musicians. Both Mackenzie and Irving became involved in the affairs of the RI. Mackenzie gave afternoon lectures and on 1 February 1895 Irving gave a Friday afternoon lecture on 'Acting: an art'. This lecture was one that would normally have been given as a Friday Evening Discourse, and which was therefore accorded publication in the *Proceedings*, but it was presumably given in the afternoon so that Irving could appear on the stage in the evening.[11] It nevertheless attracted so large an audience that over 500 people had to be turned away. On one occasion Dewar invited Ellen Terry to the RI but she was too busy to come – 'Mr Irving is *at it* every moment of his time'.[12] Stoker had degrees in science and mathematics from Trinity College, Dublin but seems not to have become a member of the RI. He is remembered today principally for his 'Gothic' novel *Dracula* which he published in 1897.

Dewar's early upbringing in the Presbyterian Church had left little influence on his later life. Of his political views he apparently said nothing in public and there is no record of him taking part in the great discussions of the day on Irish Home Rule, the Boer War, or the powers of the House of Lords. Hamilton Dickson mentioned in one of his letters 'our Carlton Club' which implies that he was a Conservative, but neither of the London clubs of that name, the Carlton Club and the Junior Carlton Club, has any record of his membership.[13] Dickson was presumably referring to a less formal Carlton Club that was active in Cambridge from the the 1890s, and maybe earlier. Dewar generally took a more progressive stance in university politics, voting, for example, with Liveing and Rayleigh for the admission of women to university examinations and degrees.[14]

For obvious reasons Dewar had few scientific friends but the most loyal, Henry Armstrong, collected some of the kinder references to his character when it came to preparing an obituary for the Royal Institution and for the Royal Society.[15] John McKendrick, Dewar's old partner in their physiological experiments at Edinburgh and later Professor of Physiology at the RI for three years, wrote to Armstrong:

> Dewar was a man of good taste. He loved beautiful things which everyone saw who was privileged to enjoy his intimate acquaintance. He took a great delight in music and especially in opera. Many a discussion I had with him on the province of opera and its place in the musical world. He was fond of poetry, more especially the poets who were very unlike himself and his thoughts. I remember once a distinguished lady complained of his coarseness and briskness of manner and she was surprised when I told her he was fond of poetry and more especially when I told her that his favourite poets were Shelley, Keats and such like. He could quote from these line upon line.[16]

Henry Savage Landor had a high opinion of Dewar's place in the scientific pantheon. After the award of Dewar's knighthood he complained of the absurdity of 'the greatest scientist in the world and the manager of the Hippodrome as having the identical award'. On the wine at lunch, Landor wrote that it was always of the finest, but: 'He and I always had to finish the bottle, which was rather more than I wanted but there was no arguing with him.'[17]

Perhaps the last word should be that of Hamilton Dickson:

> In appearance blond, in profile – with his short pointed beard – not unlike some Shakespeare bust, keen-eyed, alert in judgement, quick to act and sure of himself; pungent for the truth, fierce at the least deviation therefrom, gifted with a subtle divination of anything underhand lurking behind the most plausible presentation – one who spotted the cloven hoof almost by instinct and was merciless.[18]

His attitude to his assistants was summarised by McKendrick: 'He was severe with his assistants and not lenient with their faults, but he was a kind friend all the same to them, and willing at all times to advance their interests.'[16] This judgement was correct only to the point where he broke with them over some fault of theirs, when, as Dickson put it, he was 'merciless'. His conduct towards Ruhemann and Scott is impossible to defend, but when all went well he was respected and admired by the 'working staff', as he called them. His last senior assistant was William Green who spoke to the Library Circle Meeting in 1956. He had worked at first mainly in the Davy Faraday Laboratory from 1900, and it would be interesting to have had his side of the story of the dismissal of Scott, whom he praised warmly. He contented himself with noting that 'Contacts between the Davy Faraday and the Royal Institution became more noticeable about this time, and more jobs came my way in connection with the Professor's lectures'.[19] But he said nothing of any paid research or consulting work in which he may have participated. His almost exact contempory at the RI, Ralph Cory, who rose to be Librarian, is more explicit in his general account of Dewar's character but also avoided discussion of his outside interests:

> Dewar has been variously described by his biographers. He certainly was inclined to be irascible and excitable but he had a magnetic personality and there was something about him which far outweighed his frequent petulant outbursts. His one over-riding obsession was the Royal Institution, and for him it mattered not who or what went to the wall so long as his beloved Institution survived. I always found him ready to listen to any reasonable argument, but he would not tolerate the slightest familiarity. When he received his knighthood, on meeting him in the Library I congratulated him as 'Sir James' and was promptly told, 'Not so much of the Sir, Cory'. Incidently, I was given a crisp £5 note – enormous largesse in those days – with which to celebrate his honour in due and fitting manner. Dewar had a habit of humming – more or less tunefully – as he went about the building, which gave the staff ample warning of his approach. Promptly every morning at half-past ten, we would hear the door of his flat shut and then his humming growing louder as he came dowstairs. His first visit was invariably to the office to discuss with Young any correspondence and affairs generally. Dewar certainly kept a finger on the pulse of the Institution and knew exactly what was happening in every department.[20]

His frequent generosity to his staff and to aspiring musicians was, as we have seen, more than matched by the way in which he subsidised his expensive research programme from his own pocket. The money he received with the Rumford Medal of the Royal Society was passed directly to the RI, and that of the Gunning Victoria Prize of the Royal Society of Edinburgh went in 1902 to the newly established Tait Memorial Fund for the support of students of physics.[21] At the height of his problems over the dismissal of Scott he was paying for the RI to redecorate the Lecture Theatre to mark his holding of the Fullerian chair for as long as Faraday had held it.

His regard for Cambridge never reached the height of his concern for the RI, although he did contribute to the discussion of the design of the new chemistry department in 1885.[22] After the Chemical Club had been formed in 1901, by a group in which the Jacksonian Demonstrators were prominent, he spoke to the Club once only, on the different properties of liquid nitrogen and oxygen.[23] He was claimed as a benefactor by Peterhouse, which also houses his portrait by William Orchardson.[24] He might, however, have been given more to the College 'had we been wiser', as Dickson rather enigmatically put it, but the College papers are silent on any major row.[25] In the first decade of the twentieth century he found College and University affairs an increasing burden, as he made clear to his chemistry colleague, George Liveing (see Chapter 5). Eventually the point was reached, after the War, when the Cambridge scientists had had enough of his casual, but no doubt legally correct, attitude to his university duties. They started to look for a way of forcing him out so that they could find a Jacksonian Professor who would be of more value to the University.[26] By 1917 his salary was £600, which would have been £800 had he not been a College Fellow.[27] He died before these plans for his succession had made any progress.

The membership of the RI started to fall in 1904 as Dewar's 'magnetic personality' became less of an attraction to his London audience, and as science was displaced by the arts as a popular feature of London's cultural life.[28] His original research declined in quality and quantity after the collapse of his attempt on the liquefaction of helium in 1908. He suffered from the loss of assistants in London and in Cambridge, after the resignation of Lennox in 1908, the dismissal of Scott in 1911, and the deaths of Spivey in a laboratory accident in Cambridge in 1901 and of Jones from a fall in the Alps in 1912. The apparatus for gas liquefaction was maintained by Heath and Green, to the best of their abilities, and was in use as late as 1916 when it was reported that the vibrations of the pumps of the liquid air apparatus had caused the bursting of one of Dewar's long-lived soap bubbles; the liquid air was needed by J.J. Thomson for a Saturday afternoon lecture. But after Lennox's departure it was this apparatus that gave more trouble than the apparatus for liquid hydrogen and eventually it became unusable.[29] Shortly after Dewar's death Murray, now the managing director of the British Oxygen Company, seeing from the Annual Report of the RI that the liquid air plant was in a bad way, wrote to offer a gift of 'several gallons a week of liquid air'; one can only speculate what Dewar's response might have been![30] The best of the work at Cambridge was that of Jones on the metal carbonyls (Chapter 4) but it is unlikely Dewar contributed much to its success beyond occasional advice.

Dewar had long had a reputation of failing in his relations with his assistants but he had generally kept on reasonably good terms with his equals, with the exception of the enmity between him and Ramsay. Now he began to argue with everyone and the stresses in the RI caused by the outbreak of war in 1914 soon made matters worse. The barrister John William Gordon had become a member of the RI in 1899 and served on the Board of Visitors from 1909 to 1911. We saw in the last chapter that his role as a former Visitor led to his involvement in the final stages of the legal arguments over Scott's dismissal, and now turn to his part in a row between Dewar and William Crookes and his son Henry.[31]

Dewar's long part in the testing and analysis of water culminated in 1905 when he and William Crookes set up the Water Supply Company, of which Henry was formally an employee. On his own initiative, Henry found out how to prepare stable colloidal suspensions of metals such as mercury, gold and platinum which had valuable bacteriological properties and with which he proposed to treat water. He set up a business, Crookes' Collosols Company, to exploit this effect in which his father, after some initial worries, subscribed for some shares but took no part in the management. Dewar was outraged. He thought that Henry's role in the new company was not compatible with that of an employee of the Water Supply Company. At the least he thought that he, or the RI, was entitled to half of any profits that might be made. Henry had previously helped Dewar in preparing bacterial cultures at the RI.[32] Dewar brought a case in the Chancery Court against both Henry and William, although William was a friend and colleague of long standing and, moreover, had been the Secretary of the RI since 1900. The other officers and staff at the RI knew nothing of this row until the Assistant Secretary, Henry Young, wrote to the Treasurer, James Crighton Browne, on 7 December 1912 to say: 'I was asked yesterday [a Friday] to call on Sir William Crookes, when he informed me that he had received a writ from Professor Sir James Dewar, and expressed surprise that I was not aware that there had been any serious dispute between them.'[33] This matter apparently led to some row at a meeting on the Saturday since Dewar wrote in some anger to his friend Armstrong:

> Dear Armstrong, [7?]
> Decr 1912
>
> After the Dukes dramatic stroke today the facts will come out of what is in the background. I *assume* the resignation of Crookes and probably of the Duke. The real background is that I have lodged an Injunction in Court substantially against the Crookes Family, on the ground of Breach of Contract. Quite unknown to me H. Crookes took out Patents and formed a Company 'Crookes Collosols' of which his Father is a shareholder along with some of the Crookes relations and exploited the invention by [giving?] special and exclusive licence to same firm to sell Ag and Hg colloids for 60/- [£3] a gallon.[34]

The position was then complicated when Henry Crookes threatened Dewar with a writ of libel arising from Dewar's correspondence with the President. William Crookes dissociated himself from this threat. Dewar wrote to Crighton Browne to ask what the Managers proposed to do about the legal actions. This letter was reported to the Managers at their meeting on 14 April 1913, a meeting at which Dewar was not present. They quickly sought Crookes's resignation and a week later had installed the electrical engineer Alexander Siemens, a third cousin of Charles Siemens, as the new Secretary. He and his company had long been supporters of Dewar's cryogenic work. On 5 May, with Dewar now present again, it was resolved to ask Crookes to remain as a Manager.[35] At the end of the year, however, Crookes wrote to Siemens to tell him that he and his wife, Ellen, were resigning as Members. Crookes soon changed his mind, at the request of the President, since his resignation might be 'erroneously interpreted by scientific men in general who – ignorant of the special reasons prompting my resignation – will put on it a construction detrimental to the welfare of the Royal Institution'. In March he confirmed to Young that his wife's resignation stood; the whole episode had worried her greatly and caused her to suffer from what her husband termed 'Dewaritis'.[35]

Crookes had been at a loss of how to deal with Dewar and had earlier written to Meldola:

> I cannot make out what he really wants, and my solicitor and Counsel are in the same difficulty. From all I can make out my son has in no way broken his agreement, and if he had unwittingly done so the matter in dispute is so trifling and unimportant a character that were D a sensible man we could have settled it in five minutes. But what can I do with a man whose temper is of such ungovernable violence that one cannot speak to him without the risk of an insult.[36]

The Counsel of whom Crookes writes was none other than Gordon![37]

Dewar's case was dismissed in the Chancery Court, to the great relief of Crookes who was hoping to be elected as the next President of the Royal Society and could not afford to be involved in any scandal – his association with spiritualism was itself enough of a handicap. Dewar argued in his private correspondence with Armstrong that Larmor should be chosen.[38] In early 1913 he was writing to Armstrong, sometimes every few days, with a flow of letters that showed how obsessed he had become about what he perceived as the treachery of William and Henry Crookes. But Dewar was not a man whose opinion now commanded much respect among his colleagues and, despite this unwelcome dispute, Crookes became President of the Royal Society at the Anniversary Meeting in November 1913. He was then 81 and served for only two years, being followed by J.J. Thomson in 1915, thus continuing a sequence of five-year presidencies that has continued to this day. Crookes died in 1919; Dewar did not attend the funeral at which the RI was represented by Boverton Redwood.

In 1914 Gordon had taken up the subject of Crookes's resignation as Secretary of the RI. On 22 April he wrote to the President to complain of the the lack of respect accorded to Crookes after he was relieved of his long service: 'the fact being that Sir William Crookes, when asked to resign for the satisfaction of Sir James Dewar, definitely refused to do so, and remitted the whole question to the Managers, who, in the exercise of their discretion, relieved him of his office'. Six months later Gordon again returned to the subject of Dewar's 'removal of Crookes' by adding that two British subjects had now been replaced in the RI by two of 'German extraction'. He presumably had in mind Alexander Siemens and Robert Mond, since Ludwig was dead, and notwithstanding the fact that Robert was British-born. Alexander Siemens, he said, is 'a gentleman with whom, for a time at least, Sir James Dewar might be expected to work in amity'. He continued: 'I know – as your Grace also knows – that Sir James Dewar is a man of quarrelsome disposition and ungovernable temper, and am quite prepared to believe that these defects – together with a plentiful lack of common sense – have led him to place himself and the Royal Institution in what is, at this moment, a thoroughly false position.'[35] So far, so good, his position in regard to Crookes's resignation was a fair one, if we except his argument about the patriotism of Mond and Siemens. But now he became further distracted by the anti-German hysteria that infected so many otherwise sensible men after the outbreak of the War on 4 August 1914. He asked the President whether, at the bidding presumably of these men of German extraction, the laboratories of the RI or DFRL were 'fitted up for wireless or other signalling' or 'being used, or reserved for use, as arsenals, or otherwise, in the interests of the enemy'. The Managers took this absurd claim seriously and called in an Inspector of Police who searched the whole building and solemnly reported that he could find no evidence of such treachery.[35] Siemens himself soon came under pressure because of his 'German extraction'. He explained that he fully supported Britain's position in the war, that he had been born in Hanover and had automatically become a Prussian citizen when Hanover was annexed to Prussia in 1866, that he had moved to Britain and had now been a British citizen for 36 years. Nevertheless he decided not to seek nomination again in March 1915 and was succeeded by the patriotically irreproachable Colonel Edmond Hills, C.M.G.[39]

Dewar's relations with the Royal Society were not always harmonious. In the 1890s there was anonymous criticism of the administration of the Society that Robert Strutt thought came from Dewar, and maybe Armstrong, who thought that they should be on the Council. There is, however, no direct evidence that Dewar was involved and anonymous tactics do not seem to be his style. In 1915 his name, and that of three others, was put forward for the Society's senior award, the Copley Medal, when Crookes was the President. Meldola was the proposer and Crookes himself favoured Dewar in spite of their recent row. But when the names were discussed further in 1916, under J.J. Thomson's presidency, Ramsay and Dewar were the two candidates considered. Whether Ramsay's name was added because it was known that he was seriously ill and might not have long to live, or whether it was a proposal from those who had taken Ramsay's side in the long-running emnity between the two men, is not clear. When Ramsay died on 23 July 1916, of a cancer of the throat, there was no further opposition and Dewar received the medal at the Anniversary Meeting in November.[40]

The next year, when Dewar was on the Council, he became embroiled in a petty dispute with the President, now his RI colleague J.J. Thomson. Dewar alleged that some Fellows had deliberately changed a comma so as to alter the meaning in a report on a statement of 1875 on elections held under Statute 12, which governed the election of an occasional Fellow for his general contribution to the support of science. Thomson said that Dewar was wrong and asked him to apologise to the Fellows he had accused. Dewar would not, and wrote to Thomson to ask which Statute gave the President the authority to send this letter asking for the apology. Council decided that the President did not need any specific authority. Dewar still did not apologise. Thomson thought that the argument was not worth the trouble of further correspondence, and Council decided to let the letters 'lie on the table'. Dewar returned by asking what was meant by 'lie on the table', but Council had had enough of this dispute and made no further reply. In this storm in a tea-cup Thomson had had direct experience of the 'treacle' in which anyone who tried to argue with Dewar became embedded, as Rayleigh had found in the controversy over the discovery of argon (Chapter 9).[41] At the Anniversary Meeting in November that year Armstrong tried to oppose the re-election of Arthur Schuster as a Secretary on the grounds of his German descent. Thomson thought that Dewar was the only Fellow to support Armstrong, and the attempt failed.[42]

The role of the RI during the War was not a distinguished one. In 1914 the President, the Duke of Northumberland, was 68 and had had a stroke, the most active of the officers was the Treasurer, Crichton Browne, a psychiatrist, who was 74, and the Secretary changed from Siemens to Hills in 1915. Dewar himself was 72 and becoming increasingly difficult to deal with, even by his own exacting standards. His colleague J.J. Thomson worked mainly in Cambridge. It was not a team that was well-equipped for the wartime world and the new problems that it generated. It was again that gadfly Gordon who raised the question of what the RI was doing and what more it might do to help the war effort. In May 1915 he was working temporarily at the War Office when they applied to use the facilities at the RI to test optical equipment. The reply was a copy of the usual application form asking for admission to the DFRL as a regular worker, a reply that annoyed Colonel Turner. He was advised by Gordon, who wrote to the Managers; his letter was discussed by them at their meeting of 3 May 1915.[43] He followed it up by letters to a newspaper, *The Standard.* At a General Meeting of Members four days later Crichton Browne rebutted Gordon's claims; his report was printed and circulated for the information of members who could not attend.[44] The matter was raised in Parliament, and the War Office was able to find laboratory space at the Science Museum, but not before a letter from the President to Lord Kitchener, the Secretary of State for War, complaining of the 'impertinent' tone used by Turner, who might have been misled by 'a certain Mr. Gordon – who is, I believe, temporarily employed by the War Office in some way'. He told Kitchener that the

RI was 'at all times anxious to assist the War Office in any way in their power, which however they are assured by their legal advisers is of a strictly limited character under an Act of Parliament and their Charter'.[45] So when the War Trades Committee of the Royal Society asked for the seven to eight litres of ethylene dibromide which it believed was at the RI, Dewar handed it over.[46] He was, however, not so cooperative the next year when the Comptroller of Munitions Inventions asked if the War Office could use the RI laboratories and equipment for the study of the synthesis of ammonia from nitrogen and hydrogen at high pressures. What precisely, he asked, could the RI contribute that was not already described in seven papers and 30 or more patents by Fritz Haber? The Managers backed Dewar's view when he told the President that 'This fatuous Research cannot help H.M.'s Government in the conduct of the war'. Dewar was clearly right but the President thought that the decision not to help would look bad to the public, and he asked the Ministry to send a man round to discuss the matter with a RI sub-committee. The Ministry soon withdrew the request.[47] Dewar was no more helpful when the Ministry next asked if the RI had any staff of 'chemists with work experience'. Dewar told them that 'During my time I have had three assistants; the first passed long ago out of Science of any kind. The Remaining two both lost an eye in the course of Low temperature Investigation; one of them resigned many years ago and took, I believe, to interests connected with Engineering of some kind; the other remains here still as the General Lecture Assistant'. So he had no one available.[48] Towards the end of the war Dewar was more helpful with advice on the absorption of gas by charcoal to counter the effects of poison gases, and with the supply of liquid oxygen. But when the Naval Airship Station wanted to purify hydrogen by liquefying the impurities Dewar told them, rather pedantically, that there was no such apparatus in the DFRL, and that his own apparatus in the RI was now 'antiquated and inefficient' and that 'there were better installations for liquefying air elsewhere in the country'.[49] Nevertheless the War Office and other government departments wrote thanking the RI for its help on several ocasions. None of this satisfied Gordon who complained directly to the King that the RI was not playing its part and that it would not use its funds to buy War Loan. The Secretary told the War Office that Gordon should put his complaints directly to the members.[50] A misunderstanding of a different kind arose when the electronic firm Cossor supplied some electronic valves in which the high vacua had been produced by means of charcoal and liquid air. When the RI asked for the bill, they refused to charge since Dewar had been so helpful to them. Dewar told them that this was not appropriate since the work was not done for Cossor's benefit. Eventually the valves were returned in good condition and all seemed well but Cossor got the impression that Dewar was unwilling to continue war work, and his assistant got his call-up papers, although Dewar got this outcome reversed after the Managers told Cossor that they were mistaken over Dewar's attitude.[51]

A gloomy assessment of Dewar's state of health and mind at the end of the war was given in a letter from Young to Wilhelm Gluud, a young German chemist who had spent the year 1912 to 1913 in the DFRL:

> I regret to say that the anxiety about conflagration of the whole of the Royal Institution buildings during the Air Raids, and the near approach to success in the matter, has made him age rapidly; and he has had several break-downs in health, but he has so far managed to give a Friday Discourse each year during the war. He had no pupils at Cambridge, and the Davy-Faraday Laboratory was practicaly closed, so that but for his own work, the scientific output has been small. The Royal Institution's financial resources have been severely affected by the deaths of many old members and the support for science from outside has not been the same as in former days. The Professor has had no obligation as regards war Service, not being a Member of any Government Committee.[52]

The fall in the income of the RI was relieved somewhat when Sir Charles Parsons handed over an unexpected cheque for £1500 in April 1915 to compensate for the shortfall in membership fees. Young's reference to Dewar's own work during the war was to soap films. This subject, so remote from his cryogenic or previous chemical work seems an unlikely one for him to take up, although he had earlier shown an interest in the field, chosing it for the RI Christmas Lectures in 1878–9 and advising Rayleigh when he was giving Friday Evening Discourses on films in 1890.[53] Dewar said the main advantage of this field was that it was cheap, but its deeper aspects were not really suited to his talents. He found that if he paid scrupulous attention to cleanliness and the avoidance of airborne particles he could make soap bubbles of great size, of the order of a metre, with a surprisingly long life, of the order of weeks. As a film became thinner a boundary appeared between a thicker portion, to which the interference of light gave the familiar bright colours, and the so-called 'black film' whose thickness was less than the wavelength of light and which therefore showed no colours. The position of the sharp boundary between the two was an indicator of the direction of the horizontal plane, a fact in which he tried to interest the Air Force for fixing the direction of the local horizon in an aeroplane in flight.[54] It is not surprising that they settled on a more robust device. His manipulative skill was apparent in the making and studying of films, but the understanding of them was beyond the limit of his theoretical skill. It was a subject that physicists such as Gibbs, van der Waals and his colleague Rayleigh had developed in the 1890s but Dewar added nothing to this side of the field.[54] He did, however, train a young assistant, Stuart Lawrence, who became a leader in the experimental work on films, liquid crystals and the problems of lubrication, at Imperial College and later at Sheffield. Lawrence, who also helped William Green, Dewar's lecture assistant, in work on films until at least 1930, was the nearest Dewar came to having a research student at the RI, in the sense that he continued in the field to which Dewar had introduced him. If Davy's greatest discovery was Michael Faraday, then Lawrence's career was perhaps the most important consequence of Dewar's otherwise trivial work on films.[53]

Perhaps the last professional argument that Dewar engaged in was over a suggestion put to him by John McLennan of Toronto. He told Dewar in August 1921 that he had been in discussion with Kamerlingh Onnes, who had floated the idea that Leiden should receive some further international recognition as the leading cryogenic centre, beyond that from the Association Internationale du Froid in 1909–10. Dewar and Kamerlingh Onnes had always remained on friendly terms and it is surprising that they never met. Dewar had declined to invite Kamerlingh Onnes to come to London in 1895 to see Dewar's laboratory but, with his much more open ways, Kamerlingh Onnes would surely have been glad to have seen Dewar in Leiden; but he never came. Kamerlingh Onnes's only visit to London was shortly after Dewar's death and just before his own. Dewar gave McLennan a cold reply to the Leiden proposal: 'The idea of internationalising such research at one university in Holland, would be bad for the progress of original research, and could only emanate from an audacious scientific autonomy. I am personally unacquainted with Professor Kamerlingh Onnes and have never visited, and, in the majority of cases, never studied, the numerous communications issuing from his laboratory.' The correspondence continued with Dewar becoming increasingly testy and McLennan trying hard to maintain civil discourse. Dewar was probably right in his stance but, as on so many occasions, could have expressed his disagreement less offensively.[55]

It is of interest to see how Dewar himself described the fruits of his own career. In his last entry in *Who's Who*, he wrote that his investigations had included the following:

> Hydrogenium, Pyridine, and Chinoline bases; Physiological Action of Light, Problems of Spectroscopy; Liquefaction of Gases; the scientific use of liquid Oxygen, Air, Fluorine, Hydrogen – by evaporation of liquid hydrogen under diminished pressure, a temperature of 13° absolute or 470° F of frost was attained; at this temperature hydrogen became a frozen foam-like mass, air a rigid inert solid; only helium of known gases remained uncondensed; pure metals cooled to this temperature became almost perfect conductors of electricity. By the use of charcoal as an absorbing agent for gases at low temperature the highest vacuum can be rapidly obtained, and gases like hydrogen, helium, and neon separated from air. Problems of capillarity; soap films of the greatest tenuity like the black of Newton's rings, which is about a hundred-thousandth of a millimetre thick, were shown to be exceedingly stable and long-lived, when made in a cryophorous vacuum; soap bubbles when blown in perfectly pure air can be preserved for a period of two months or more, and can reach a thinness when the bubbles are uniformly black, having lost all colour; photographs taken in natural colours of bubbles 42 cm. diameter (17 inches), in various stages of development, were exhibited at the Royal society, 22 June 1916.

This list is remarkable in several ways; it opens with a non-existent substance, it lays little emphasis on his greatest achievement, the liquefaction of hydrogen, and it exhibits a common weakness of elderly scientists by spending so much space on their latest work, greatly overestimating its importance.[56] It omits also any mention of another of his last interests, the study of the thermal radiation from the night sky. The physics and chemistry of the atmosphere had long interested him. In 1888–9 he had chosen the subject 'Cloud and Cloudland' for the RI Christmas Lectures, and he gave Friday Evening Discourses on 'Problems of the atmosphere' and 'Cloudland studies' in 1902 and 1921 respectively.[57] It was this last work, in which he used his charcoal thermoscope to measure the radiation from the night sky, that probably led to his death. On Monday 19 March 1923 Ralph Cory, then the Library Assistant, was about to leave the building when he was handed a telegram by the Hall Porter. It was addressed to Dewar and was from Hendrik Lorentz at Leiden who was scheduled to give a Friday Evening Discourse on 1 June on 'The radiation of light'. Cory was about to put it in his pocket and deliver it to Dewar the next morning when he thought that the reaction would be 'Why wasn't I shown this last night?'. So he went up to the flat and found Dewar and his wife in front of the fire – it was a cold night – with a tea-table between them. Dewar began to dictate a reply but then broke off, saying 'No, it's getting late – we cannot do anything about it tonight, see me in the morning – get along home – good night'. Later he went to the laboratory and then on to the roof to study again the radiation from the cold air. In the early hours he had a severe stroke and was put to bed. He never recovered and died on Tuesday, 27 March.[20] J.D. van der Waals, a man he held in great respect, and by then the senior of the Honorary Foreign Members of the Royal Institution, had died three weeks earlier, on 8 March. A few months later, on 2 August, Robert Lennox decided that, with Dewar now no longer there, the time had come for him to become a member of the RI.

Cremation had been legal de facto since 1885 when a prosecution had failed, but was formally legalised only in the Cremation Act of 1902. Dewar had been a member of the Cremation Society for many years and had made it clear in his will that he desired that 'the funeral arrangements made on my death may be of a simple character and kept entirely private and confined to the members of my family and that no religious service be held at the Chapel attached to the Crematorium or elsewhere'. He asked that there should be no book-length biography and no commemorative prize or fellowship.[58] The cremation took place on the Saturday, 31 March, four days after his death. A small group of friends assembled in the morning at the Royal Institution; the names of these are not known precisely but they probably included the President of the RI, the Duke of Northumberland, Alexander Mackenzie from the Royal Academy of Music, and certainly included the Treasurer, James Crichton Browne, and the Secretary, Arthur Keith with his wife Celia, since all these (except for Celia) were at the cremation itself at the Golders Green Crematorium later in the day, with another friend, Arthur Landor, the traveller. They were surprised that morning to find that the Bishop of Worcester was at the RI conducting a short service. Apparently Crichton Browne had decided that the wishes of the living took precedence over those of the deceased. Throughout the accounts of these proceedings there is no mention of Helen Dewar; perhaps, following Scottish custom, she had decided that a widow should not attend her late husband's funeral. Dewar's ashes were placed in a large and elaborate marble casket, which is still displayed at the crematorium alongside a similar one containing the ashes of Helen Dewar, put there after her death in 1935. Cory says that a service for the staff at the RI was conducted in the study by Dewar's 'old friend' Canon Carnegie, but this may have been on a later day.

Dewar died a wealthy man; the net value of his estate being assessed at £128288. His will of July 1922, which may not have been his first, was long and complicated.[58] All his brothers had died before him and he appointed his wife and his nephew, Dr Thomas William Dewar of Leith, as joint executors and trustees. He left to his wife his personal and household possessions, £25000 and an annuity of £100. He left to each of his assistants who had been at the RI since 1900 gifts not exceeding £500, at the discretion of Hamilton Dickson. He left £500 each to Dickson, to Munro Ross and to his nephew Ernest Scott Dickson (son of his late brother-in-law, Charles Scott Dickson who had died the previous year) to help his wife in an edition of his papers similar in form to that of his spectroscopic papers in 1915. His scientific equipment was to go the RI and to Cambridge University. The rest of his estate was to be placed in trust to provide an income for his wife during her lifetime and was then to be distributed, as follows; £5000 to the Royal Institution to add to the endowment of the Fullerian Professorship, £1500 each to his nephews Thomas William Dewar and James Campbell Dewar, £1500 each to Mrs Dorothy George Lammie of Edinburgh and to Mrs Elizabeth Stoddart (sisters of his wife). His charitable donations then followed: £5000 to the Royal Institution, £3000 to the Royal Academy of Music, £3000 to Dollar Institution, £2000 for the poor of Kincardine, and any remainder to such charitable bodies in Scotland as the executors may choose. There was no mention of Peterhouse, an absence which might account for Hamilton Dickson's remark in his obituary that the college might have done better 'had we been wiser'.

This will, however, survived unchanged for only five months. In a codicil of 18 December 1922 he made his wife the sole executrix and rescinded the donations to charitable institutions.[58] There seems to have been no row with Thomas William Dewar, the deposed executor, since he was given a role in the handling of the estate if Helen Rose Dewar should die before him. In the more likely event of her survival she was asked to take the advice of his solicitor James Curzon Mander 'in making any testamentary disposition of the estate she will derive from me … as he is well conversant with my private affairs and has knowledge of the general direction of my wishes which have changed from time to time'. She died on 7 January 1935, and was cremated at Golders Green on 11 January. She left £106508, so she had handled her legacy frugally.[59] Her estate included the fine French furniture and the paintings, sketches and engravings that had adorned the walls of their flat at the RI. The dispositions she made were not dissimilar to those of James's original will: £30000 to the Royal Institution with other substantial gifts to the Royal Academy of Music, to the Royal Society Mond Laboratory and to Peterhouse at Cambridge, and to the poor of Kincardine, but with the residue to go to the four Scottish universities and to the Royal Infirmaries of Edinburgh and Glasgow.[59] The RI decided to use some of their share to set up a Dewar Fellowship, to which Paul Ubbelohde, of Christ Church, Oxford, was appointed in 1936.[60] He held the post for four years before he was swept into the scientific civil service after the outbreak of the Second World War.

A curiosity of James's original will is that he felt obliged to say publicly where his wealth had come from. He wrote: 'I desire to explain that in making the above provisions in this my will I have taken into account the fact that practically the whole of the revenue received by me from my Professorships and technical scientific professional labours has been devoted to the execution of my physical and chemical investigations and that the accumulated funds were chiefly derived through a bequest from a brother.' All his brothers were dead when this will was made so it is not clear why he deliberately refrains from mentioning which brother left him a bequest. Perhaps the most likely was the 5th brother, Hugh, who lived in Melbourne, since he was the one who supported the RI experiments with gifts of £100 in 1893 and £200 in 1898.[61] Nevertheless one has to suspect some exaggeration here since Dewar's expensive style of living at the RI seems to have started early in his career and to have been supported by his considerable non-professorial earnings.

In his last years and after his death Dewar was thought of more kindly than was often the case in his most active days. He and his wife were given a silver-gilt loving-cup by members of the RI and others on the occasion of the jubilee of their wedding in 1921. Subscribers were asked to give not more than a guinea and no fewer than 273 responded.[62] This number reflected perhaps the universal admiration of his wife's skill as a host as much as admiration of his skill in research and as a lecturer. The cup was presented by the President after a Friday Evening Discourse by J.J. Thomson on 17 June 1921. After his death the question arose of the form a memorial to him should take, in contravention to his own wish for none. After much discussion, a bronze relief mounted in a marble surround, by the Australian sculptor Bernard Mackennal was chosen. It was unveiled by Munro Ross on 12 November 1925 and was for many years displayed in a niche above the first flight of stairs in the entrance hall.[63] It was removed and placed in storage in 2006 when the niche was extended into a window from the hall into the corridor behind.

The formal tribute of the RI to its longest-serving professor took place at the General Monthly Meeting of the members on 9 April 1923. The Secretary, Arthur Keith, read a long account of all that Dewar had accomplished and of his services to the Royal Institution; the President, the Duke of Northumberland, moved a motion approving of this tribute, which was seconded by the Treasurer, James Crichton Browne. The last was the one who had known Dewar best for more than 50 years, and was one of the rare colleagues with whom he had been on good terms throughout their acquaintanceship. It was appropriate that, on such an occasion, he could expresss all that was most admirable in Dewar's character:

> He had a charming and approachable manner which won him a host of friends. He had a very attractive personality. Bright, breezy, open-hearted and not without the saving grace of humour he was one of the most lovable of men,

> always accessible to the student, full of encouragement to the honest worker, brimful of knowledge and as generous with his brains as with his purse.

He was followed by shorter tributes from Rutherford, J.J. Thomson and Armstrong. The less admirable side of his character was glossed as his hatred of all that was meretricious or dishonest in research, and his shortcomings in the theoretical understanding of modern physics were disguised by justifiable tributes to his experimental skill. The report of the meeting includes lists of the lectures that Dewar gave at the RI: 49 Friday Evening Discourses and one Commemorative Lecture in 1899, and 39 courses of day lectures, of which eight were the Christmas Lectures for children.[64]

Dewar himself had hoped that he would be followed as Director by Rutherford who had succeeded Thomson as the part-time Professor of Natural Philosophy at the RI in 1921. He was, like Dewar, an avowed experimentalist who left the deeper aspects of theory to his colleagues, and who had already agreed to step in as acting Director of the DFRL. So the Treasurer and Secretary went to Cambridge to see if he could be persuaded to take on the larger role. He declined, however, and suggested instead that they invite Bragg. The RI Officers took him to mean Lawrence, the younger Bragg, then Professor of Physics at Manchester, but Rutherford said, 'No – I mean Sir William Bragg, the Professor of Physics at University College; he is a great man of science, and he is also a very great man'. Perhaps to their surprise, William agreed and soon revived the RI itself and, in effect, united it with the DFRL. It now became a leading centre for the study of molecular structures by X-ray diffraction under first William and later Lawrence Bragg, and with the support of the able workers that they attracted. The 20 years of decline were over and the Royal Institution again flourished.[65]

Chronology

1842 September	Birth of James Dewar
1852 March	Death of his mother
1853–4	James suffers from injuries after a fall through the ice
1857 September	Death of his father
1858 September	Enters Dollar Institution
1859 September	Enters Edinburgh University
1867	First paper, published in *PRSE*
1867 Summer	Works with Kekulé in Ghent
1868 August	First paper at BAAS, and becomes a member
1869–74	Assistant to the Professor of Chemistry at Edinburgh
1869 February	Elected Fellow of the Royal Society of Edinburgh
1869 Autumn	Appointed Chemist at Dick Veterinary College
1870 December	Becomes a Fellow of the Chemical Society
1871 August	Marries Helen Rose Banks
1872 November	First Friday Evening Discourse at the Royal Institution
1873 February	First lecture to the Chemical Society, on 'Dissociation'
1873 March	Appointed Assistant Chemist, Highland and Agricultural Society of Scotland
1873 November	Member of Council of Royal Society of Edinburgh
1874–7	Member of Council of Chemical Society
1874 April	Applies for the Chair of Chemistry at Glasgow
1875 April	Elected Jacksonian Professor of Natural Experimental Philosophy at Cambridge. Alexander Scott appointed to be his assistant
1877 April	Elected Fullerian Professor of Chemistry at the Royal Institution
1877 June	Elected Fellow of the Royal Society
1877 December	Liquefaction of oxygen by Cailletet and by Pictet
1878	Starts collaboration with Liveing on spectroscopy
	Gerrard Ansdell becomes his first assistant at the Royal Institution
1879 September	President of Section B, BAAS at its meeting in Sheffield
1880–2	Vice-President of Chemical Society
1881	Founder member of the Society of Chemical Industry
1881 December	Robert Lennox succeeds Ansdell as assistant at the RI
1883	Elected a Fellow of Peterhouse, Cambridge
	Receives shares in the Nobel oil company

1884	Liquefies oxygen in bulk for the first time in Britain Elected member of the Athenaeum Club
1884 June–July	Visits Canada and USA for meetings of the BAAS and AAAS
1884 August	Proposes Alfred Nobel for election to the Royal Society Visits Nobel's oil wells at Baku
1884 November	Appointed 'chemist to Nobel's company' in Britain
1885	Applies for the chair at Royal College of Science Siegfried Ruhemann becomes Dewar's assistant in Cambridge in succession to Scott
1886 May	Accident with ethylene and liquid oxygen that incapacitates Dewar for six months
1886–7	Vice-President of Chemical Society
1887–8	President of the Society of Chemical Industry
1888 August	War Office Explosive Committee set up
1888 November	Nobel brings 'ballistite' to the Explosive Committee
1889 July	Dewar and Abel invent 'cordite'
1889 August	Cordite patented
1890 December	Dewar tells Ruhemann to resign
1891 June	Dispute with Ruhemann resolved
1891 June	Faraday Centenary celebrated at the RI
1891 Autumn	Scott returns to Cambridge as Dewar's Demonstrator
1891 October	Issues a pamphlet on the Ruhemann affair
1892	Starts collaboration with Fleming on electrical and magnetic properties at low temperatures
1892 May	Appointed to the 'Balfour Commission' for a report on the London water supply
1892 October	Hodgkins bequest notified to RI
1892 December	Vacuum flask shown at the RI
1893 September	Balfour Commission reports
1894 January	Nobel sues for the infringement of his ballistite patent
1894 July	Nobel takes his case to the Court of Appeal
1894 August	Joins Crookes on committee to analyse London water Receives the Rumford Medal of the Royal Society President of the engineering club, The Smeatonians Rayleigh announces the discovery of argon at BAAS
1894 November	Hampson proposes cooperation with the RI for the liquefaction of air by his new method
1895 February	Nobel loses his case before the House of Lords
1895 March	Ramsay announces discovery of helium in clèveite
1895 ca. June	First mention in his notebook of Dewar's 'direct' apparatus for liquefying air
1895 December	Liquefies air using the Joule-Thomson effect

1896 Autumn	Establishment of the Davy Faraday Research Laboratory by Ludwig Mond. Scott comes to London as its Director, and is replaced as Demonstrator in Cambridge by William Spivey
	Honorary LL.D. Edinburgh
1897–9	President of the Chemical Society
1898 May	Liquefies hydrogen for the first time
1898 June	Ramsay isolates neon
1899 April	Dewar receives Hodgkins Medal from Smithsonian Institution
1899 June	Celebration of the Centenary of the RI
1901 June	Delivers the Bakerian Lecture to the Royal Society
1901 October	Death of Spivey in a laboratory accident at Cambridge, and his replacement as Demonstrator by H.O. Jones
1902	President of the BAAS
1904	Knighted
	Hon. D.Sc., Oxford
	Receives Lavoisier Gold Medal of Académie des Sciences
	Receives Gunning Victoria Prize of Royal Society of Edinburgh
	Apparatus for the liquefaction of hydrogen exhibited at the St Louis Exhibition
1905 October	Metropolitan Water Board takes over the testing of London's water
1906	Receives Matteucci Medal of Societa Italiana delle Scienze
1908	Receives Albert Medal of Royal Society of Arts
1908 July	Kamerlingh Onnes liquefies helium
1908 November	Resignation of Robert Lennox
1909	Receives the Davy Medal of the Royal Society
1910	'Dismissal' of Alexander Scott from the DFRL
1912 August	Death of H.O. Jones in the Alps
	William H. Mills appointed Jacksonian Demonstrator
1916 November	Receives the Copley Medal of the Royal Society
1918 May	Receives Medal of the Society of Chemical Industry
1919	Receives Franklin Institute Medal
1923 March	Death of James Dewar
1935 January	Death of Helen Rose Dewar

Appendix
Liquefying a Gas

Simple substances exist in three possible states: gas, liquid and solid. The prevailing temperature and pressure determine in which state each substance is found, as is shown by means of a graph of pressure as a function of temperature in which the boundaries between the three phases are shown as lines (Figure A.1). The three boundaries – gas-solid, gas-liquid and liquid-solid meet at the triple point (t.p.). For most substances the pressure at this point is well below atmospheric pressure of 1.03 bar, e.g. oxygen 0.002 bar, water 0.006 bar, nitrogen 0.125 bar, and argon 0.689 bar, but for carbon dioxide it is as high as 5.18 bar. Cooling a gas at atmospheric pressure (shown schematically by the horizontal arrow at the right of Figure A.1) leads therefore to the succession of phases of gas to liquid to solid. For carbon dioxide, the gas at atmospheric pressure goes directly to the solid state and no liquid is seen unless the cooling takes place above 5.18 bar.

The boundary separating liquid and gas ends at the pressure and temperature of what is called the critical point (c.p.). As we heat a liquid in equilibrium with its vapour (or gas), the liquid becomes less dense and the gas more dense as the pressure rises. At the critical point the two assume the same state and beyond this point the state can be described only as 'fluid', without any distinction between liquid and gas. It follows, therefore, that if we wish to liquefy a gas we must first cool it to a temperature below that of the critical point. Only below this temperature is there a state of the system in which the two phases, liquid and gas, are different and separated by a visible meniscus.

Three ways of expanding a gas are set out in Section 8.4. The second is not used for gas liquefaction and the first and third are discussed here in more detail. For the first we consider particularly the form of the experiment as used by Cailletet in his original liquefaction of oxygen. Here the gas is in a cylinder under pressure and is expanded by opening a valve leading to the space outside which is at a pressure low compared with that in the cylinder. In the third, the gas flows continuously down a pipe in which there is a plug or obstruction that causes a fall in the pressure as the gas passes the plug.

For Cailletet's expansion, let us consider first that hypothetical entity, the perfect (or ideal) gas, for which the pressure, volume and temperature are linked by the equation that expresses the laws of Boyle and Charles (or Gay-Lussac), namely $pV = RT$, where the volume, V, and so the gas-constant, R, are both proportional to the amount of gas. We start with the compressed gas in the cylinder, Figure A.2, and with the valve closed. The cylinder is thermally isolated; that is, no heat can enter or leave through its walls. It is divided into two parts by

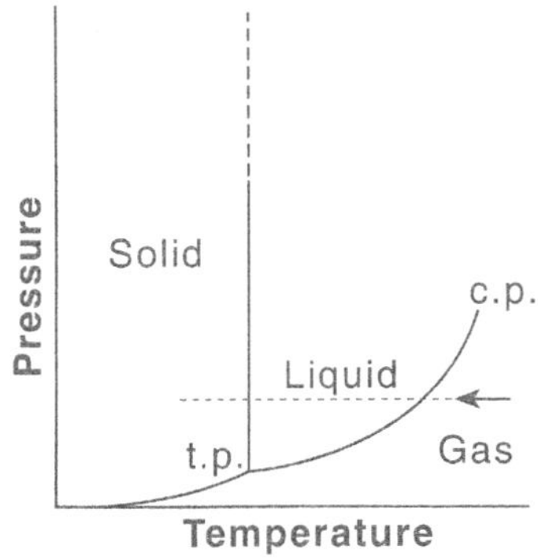

Figure A.1 The three phases of matter showing the usual effect of cooling a gas at a constant pressure, taking it first to a liquid and then to a solid. 't.p.' and 'c.p.' show the triple and critical points.

a piston which is deemed to be without mass, to move without friction, and also to be impervious to heat. It follows that the portions of gas on the two sides of the piston always have the same pressure but may differ in temperature. We consider here only the state of the gas to the left of the piston and suppose that its amount is 1 mole (i.e. 32 g for oxygen); this portion of gas is, in thermodynamic terms, our *system*. At the end of the expansion this piston will have moved to the position shown by the dotted line. We do not have to consider the state of the gas to the right of the piston or that of the gas as it leaves through the valve. The piston is introduced here only for convenience in visualising what is happening; the result is essentially unchanged if it is omitted. On opening the valve the gas on the right of the piston is blown out by that on the left which does work in expelling it. For an infinitesimal movement of the piston the volume of the confined gas increases by $\mathrm{d}V$ and the work done is $(p\,\mathrm{d}V)$. Since the gas is perfect this is $(RT/V)\,\mathrm{d}V$. Because of the thermal isolation of the gas on the left, this work can come only from its own internal energy, and, for a perfect gas, this energy resides only in the kinetic energy of the molecules and in the internal energy of each molecule. The important point here is that in a perfect gas there is no energy arising from any interaction of one molecule with another; the energy is independent of the arrangement of the molecules and so of the volume to which the gas is confined. The internal energy of the gas, U, is therefore a function only of its temperature, a result that we can express by writing $U = C(T).\ T$, where the constant of proportionality may, in general, be itself a function of temperature. (The zero from which U is measured is irrelevant.) It is here that the usage of engineers and pure scientists is sometimes different; for the former it is often convenient to the define a perfect gas as one for which C is a constant, independent of temperature, and for the latter no such restriction is imposed. However for the two 'permanent' gases that were of the greatest importance in the 1870s and 1880s, oxygen and nitrogen, C is, in fact,

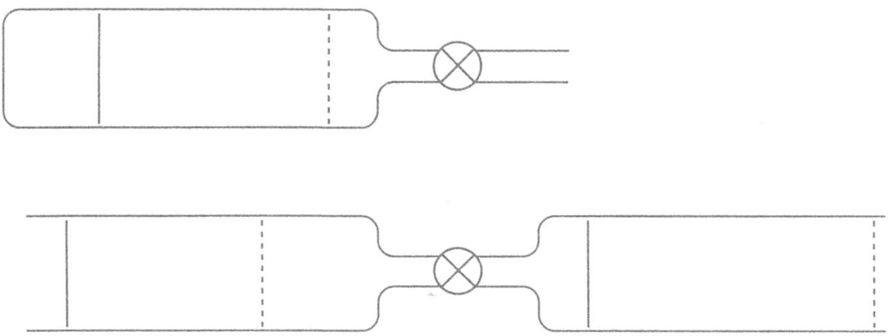

Figure A.2 Cooling by expansion in a Cailletet tube (top) and by Joule-Thomson expansion (bottom).

almost independent of temperature at low and modest pressures. For calculations of the simplicity attempted here, it may be taken to be a constant; it is $5R/2$ for one mole of each of these two gases; so $U = 5RT/2$. By equating the work done to the fall in energy, we have:

$$(RT/V)\,\mathrm{d}V = -\,\mathrm{d}U = -\,C\,\mathrm{d}T = -\,(5R/2)\,\mathrm{d}T$$

An integration of the two outer parts of this equation for an expansion from V_1 to V_2, and an application of the perfect-gas law, gives:

$$\ln (T_1/T_2) = -\,(2/5) \ln (V_1/V_2) = (2/7) \ln (p_1/p_2)$$

Since p_2 is less than p_1 it follows that the gas to the left of the piston suffers a fall in temperature. If we consider the expansion of oxygen or nitrogen from a pressure of 300 atmospheres to atmospheric pressure and from an initial temperature of 20 °C, then the calculated final temperature is 57 K on the absolute or thermodynamic scale. This result is certainly in error since the gases are not perfect at room temperature at a pressure of 300 atmospheres, nor at 57 K and at atmospheric pressure. In fact, oxygen at one atmosphere is a liquid below 90 K, but the aim of the calculation is to show only how great is the cooling and so show that this mode of expansion is thermodynamically effective, although it is mechanically difficult to put into practice in a way that produces liquid oxygen in quantity. The cooling is, moreover, universal in practice since it does not, except for the numerical details of the calculation, depend significantly on the properties of the gas under consideration. The smaller the heat capacity, C, the more effective is the process, so the method is well-adapted to liquefying the inert gases such as helium and neon.

An expansion of this kind is called adiabatic since the essence of the process is the thermal isolation of the gas, and it is reversible in thermodynamic terms since the pressure is always the same on both sides of the piston. Such an expansion is one at constant entropy. C is the heat capacity of a gas heated at constant volume, and therefore usually written C_v. A gas heated at constant pressure has a larger heat capacity, C_p, since as well as increasing its internal energy the gas must do work to increase its volume in order to maintain the pressure constant; for a perfect gas this excess is R, so $C_p = C_v + R$. It follows that for a perfect gas an isothermal change of pressure and volume are described by the statement that pV remains constant, while an adiabatic change is described by saying that pV^γ is constant, where γ is $C_p/C_v = 1 + (R/C_v)$. This second law of expansion was used by Laplace to correct Newton's calculation of the speed of sound in air by pointing out that in a sound wave the gas is subject to alternate adiabatic compressions and expansions, not to isothermal changes.

The other useful way of expanding a gas is by flowing it along a pipe in which there is a partial obstruction that induces a fall of pressure, as in the third expansion of Section 8.4 (see Figure A.2). Again we require that the pipe, the obstruction and the imagined pistons that isolate a portion of gas are all impervious to heat. The form of the obstruction is immaterial but for experiments on gas liquefaction it was again a partially open valve. Let us consider the portion of gas between the two pistons, flowing for a time that corresponds to the displacement of the left-hand piston by a volume ΔV_1 and the right-hand piston by ΔV_2; this portion of gas is now our thermodynamic system. The gas is passing continuously in a steady-state of flow so the piston on the left does work on the gas of $p_1\Delta V_1$ and the gas, after passing the valve, does work on the piston on the right of $p_2\Delta V_2$. Neither pressure changes during the flow so these expressions can be written $\Delta(pV)_1$ and $\Delta(pV)_2$. The difference between these two expressions for the work is, as before, equal to the change in the internal energy of the portion of the gas that has passed the valve, $\Delta(U_2 - U_1)$; that is, $\Delta(U + pV)$ is constant. We may now drop the Δ since this serves only to define the distance moved by the pistons, and so the amount that passes through the valve. This steady adiabatic flow is therefore described as one in which $(U + pV)$ is unchanged in the portion of gas passing the valve. It is convenient to give this sum a name and a symbol; it is the enthalpy and is represented H. It plays the role for a flowing gas that U plays for a static gas and its temperature derivative at constant pressure is C_p.

For a perfect gas both U and pV, and hence H, are independent of the volume of the gas and are functions of its temperature only. We can express the change of temperature on adiabatic flow through an obstruction by the symbol $\mu = (\partial T/\partial p)_H$, where, by the usual rules of partial differentiation;

$$\mu = (\partial T/\partial p)_H = - (\partial H/\partial p)_T/(\partial H/\partial T)_p = - (\partial H/\partial p)_T/C_p$$

If H is independent of pressure at a constant temperature it follows that μ is zero; there is no change of temperature for a perfect gas. This form of expansion is therefore fundamentally different from the adiabatic reversible expansion considered in the first case and it seems an unpromising route to cooling and so liquefying a gas. In practice it is found that μ is small but positive, around 0.20 K for nitrogen and 0.25 K for air at room temperature for a fall of pressure of 1 atmosphere. It depends on the particular properties of the gas considered. It was first measured by James Joule and William Thomson in the years 1852–62, and μ is now called the Joule-Thomson coefficient.

It is not surprising that at a time when the laws of thermodynamics were novel and little understood, particularly by chemists, that there was confusion between these two radically different ways of expanding and so cooling a gas. At first the most common errors were to assume that there was no real difference between the two expansions or to think that the Joule-Thomson coefficient was too small to have an important role in the matter. Another error was a failure to distinguish between the Joule-Thomson effect, $(\partial T/\partial p)_H$, and Joule's earlier attempt to measure the cooling by a single-shot expansion without doing any work, that is the change $(\partial T/\partial p)_U$, as set out as the second expansion of Section 8.4. More recently the fault has been to ascribe to the Joule-Thomson effect methods of cooling that are closer to the first expansion above, that is, by doing as much work as possible, $(\partial T/\partial p)_S$. Serious errors can be found in monographs and textbooks from 1899 to 1990, and still persist; the recent Wikipedia article on Cailletet (accessed 11/10/2010) wrongly describes his apparatus as an example of Joule-Thomson cooling.

Notes and References

Chapter 1 Boyhood

1 G.F. Black, *The Surnames of Scotland* (New York Public Library, 1946; reprinted Birlinn, Edinburgh, 1993).

2 See ancestry.co.uk/facts/Dewar-family-history-sct.ashx (accessed 3/11/2009).

3 The principal account of Dewar's early life is a booklet by W. Meiklejohn, *Tulliallan: Four Lads o'pairts: Sir James Wylie, Sir James Dewar, Robert Maule, J.P., Sir Robert Maule* (published privately, 1990, and available from the Kincardine Local History Group). Meiklejohn's principal source was reports in the local newspapers. Most of the unreferenced facts in this and the next chapter come from this booklet. Tulliallan is the name of the parish that includes Kincardine.

4 The Inn is still in business although it no longer offers overnight accommodation but is a restaurant and bar. It is claimed that it was built in 1639, but John Gifford, in *Fife*, in the series *Buildings of Scotland* (Penguin Books, London, 1992) makes no mention of it and casts doubt on several similar claims for erection in the seventeenth century for neighbouring buildings. Kincardine became a 'burgh of barony' only in 1663.

5 Helen Rose Dewar to Henry Armstrong, 20/5/1924, *RIA* DC 313a. 'Ann', according to a certified copy of James's birth certificate, *RIA* DB 11/13.

6 J. Crichton Browne, 'Sir James Dewar', *Sci. Progress* **18** (1923–4) 212–33.

Chapter 2 Edinburgh

1 H.E. Armstrong, *James Dewar* (Benn, London, 1924). This pamphlet is Armstrong's Friday Evening Discourse at the Royal Institution on 18 January 1924. It was never published in the *PRI* but was printed and issued free to all members. It formed the basis of his later obituaries, *PRS A* **111** (1926) xiii–xxiii, and *JCS* **131** (1928) 1066–76; *DSB*; *ODNB*. See also J. Crichton Browne, 'Sir James Dewar, L.L.D., F.R.S.', *PRSE* **43** (1922–3) 255–60, and Chapter 1, note 6. Armstrong and Crichton Browne, a medically-trained Scot, were close friends of Dewar and the source of many of the minor tales of Dewar's early professional years, but Armstrong wrongly gives the year of Dewar's marriage as 1872. The newspaper announcement confirming 1871 is with Dewar's papers at the Royal Institution, *RIA* DE 11/4/1, but the actual place of the ceremony may have been in a private house, according to Meiklejohn's quotation of a report in *The Scotsman*. For the death of Helen Rose, see 'Lady Dewar', *The Times*, 9/1/1935. For the playing on the fiddle, see A. Keith, *An Autobiography* (Watts, London, 1950) p. 442.

2 D.B. Horn, *A Short History of the University of Edinburgh, 1556–1869* (Edinburgh University Press, 1967); R.M. Birse, *Science at the University of Edinburgh, 1583–1993* (Faculty of Science and Engineering, University of Edinburgh, 1994); R.D.

Anderson, 'The construction of a modern university', Part III of R.D. Anderson, M. Lynch and N. Phillipson, eds, *The University of Edinburgh: An Illustrated History* (Edinburgh University Press, 2003); W.P. Doyle, www.chem.ed.ac.uk/public/professors/playfair.html and www.chem.ed.ac.uk/public/professors/crum-brown.html (accessed 5/11/2009).

3 P.G. Tait (1831–1901); *DSB*; *ODNB*; C.G. Knott, *The Life and Scientific Work of Peter Guthrie Tait* (CUP, 1911) p. 51.

4 J.C. Maxwell (1831–79); *DSB*; *ODNB*; [W.D. Niven], *PRS* **33** (1881–2) i–xvi. P.M. Harman, *The Natural Philosophy of James Clerk Maxwell* (CUP, 1998).

5 L. Playfair (1818–98); *DSB*; *ODNB*; A. C[rum] B[rown], *PRS* **64** (1898–9) ix–xi. [T.]W. Reid, *Memoirs and Correspondence of Lyon Playfair* (Cassell, London, 1899; reprinted by Pollack, Jemimaville, Invernesshire, 1976) pp. 180–3.

6 A.C. Brown (1838–1922); *DSB*; *ODNB*; J. W[alker], 'Alexander Crum Brown', *JCS Trans* **125** (1923) 3422–31; *PRS A* **105** (1924) i–v. Crum Brown was a brother-in-law of Guthrie Tait.

7 Hamilton Dickson to Henry Armstrong, 11/11/1924, *RIA* DB 11/1/13.

8 John Young Buchanan (1844–1925); *DSB*; *ODNB*; A.E. S[hipley], *PRS A* **110** (1926) xii–xiii.

9 H. Armstrong, 'The Dyestuffs Industry', *Nature* **135** (1935) 907; M.R. Fox, *Dye-Makers of Great Britain, 1856–1976* (ICI Ltd, Manchester, 1987) pp. 104 et seq. For Edward Charles Nicholson (1827–90), see *JCS Trans* **59** (1891) 464–5.

10 D.F. Larder, 'Alexander Crum Brown and his doctoral thesis of 1861', *Ambix* **14** (1967) 112–32.

11 A. Kekulé (1829–96); *DSB*. Memorial Lecture, note 20 below.

12 L. Playfair, 'On the constitution of salts', *JCS* **16** (1863) 274–97. A full account of the developments of the 1860s is in C.A. Russell, *The History of Valency* (Leicester University Press, 1971) chap. 5, 6 and 12.

13 A.C. Brown, *On the theory of chemical combination*, MD thesis, Edinburgh, 1861, printed and published by the author in 1879, and now available at homepages.ed.ac.uk/aa01/crum/crum-brown-md-thesis.pdf (accessed 9/11/2009); see especially pp. 17–18 of the thesis.

14 A.C. Brown, 'On the theory of isomeric compounds', *JCS* **18** (1865) 230–45. A.W. Hofmann had used similar models in his Friday Evening Discourse at the Royal Institution in the same year: 'On the combining power of atoms', *PRI* **4** (1865) 401–30.

15 A.[M.] Butlerov, 'Einiges über die chemischer Struktur der Körper', *Zeit. Chemie und Pharmacie* **4** (1861) 549–60.

16 J. Dewar, 'On the oxidation of phenyl alcohol, and a mechanical arrangement adapted to illustrate structure in non-saturated hydrocarbons', *PRSE* **6** (1867) 82–6; *CP* 1. Russell, note 12 above, p. 248, interprets 'phenyl alcohol' as benzyl alcohol but it is called phenol in Dewar's papers at the British Association at Norwich in 1868 (see following note 17). Dewar was elected to the Royal Society of Edinburgh on 15 February 1869, on the proposal of Playfair.

17 J. Dewar, 'On the coal tar bases', *Rep. BAAS Trans. Sec.* **38** (1868) 35–6, 'On Kekulé's model to illustrate graphical formulae', ibid., 36; *CP* 3 and 4. See also H. MacLeod, *Diary*, at the RI, 25/8/1868.

18 J. Dewar and H.O. Jones, 'The chemical reactions of nickel carbonyl, Part II; Reactions with aromatic hydrocarbons in the presence of aluminium chloride.

Synthesis of aldehydes and anthracene derivatives', *JCS* **85** (1904) 212–22, *CP* 215.

19 E.E. van Tammelen and S.P. Pappas, 'Chemistry of Dewar benzene. 1,2,5-Tri-t-Butyl bicyclo[2,2,0]-2,5-Diene', *J. Amer. Chem. Soc.* **84** (1962) 3789–91; 'Bicyclo[2,2,0]hexa-2,5-diene', *J. Amer. Chem. Soc.* **85** (1963) 3297–8. See also W. Baker and D.H. Rouvray, 'Para-bond or "Dewar" benzene?', *J. Chem. Educ.* **55** (1978) 645; R.H. Goldsmith, 'Origins of the so-called Dewar benzene formula', *Ambix* **27** (1980) 136–40.

20 Dewar's own brief account of his time in Ghent is appended to the Kekulé Memorial Lecture given by F.R. Japp on 15 December 1897, *PCS* **13** (1897) 235–40; *CN* **76** (1897) 323–5, when Dewar was President of the Chemical Society.

21 Brewster to Plateau (2/4/1867) and to Quetelet (3/4/1867): *RIA* D 1e/9 and 10.

22 W. Körner [G. Koerner] (1839–1925); J.B. Cohen, *JCS Trans* **127** (1925) 2973–82.

23 J. Dewar, 'On the oxidation products of picoline', *TRSE* **26** (1872) 189–96, with an abstract, read 6 June 1870, in *PRSE* **7** (1870) 192–3; *CP* 10 and 9; *CN* **23** (1871) 38–41, published 27 January 1871, and abstracted in *JCS* **9** (1871) 144–5 by C[arl] S[chorlemmer] who indicated that he already knew of Koerner's claim to this structure. Dewar's suggestion that the structure could be confirmed by synthesising pyridine from two molecules of acetylene and one of hydrogen cyanide was put into practice by William Ramsay, then an assistant at Glasgow: 'On picoline and its derivatives', *PM* **2** (1876) 269–81.

24 G. Koerner, 'Synthèse d'une base isomère à la toluidine', *CR* **68** (1869) 824.

25 The confusion surrounding the proposals for the structure of pyridine has been resolved by L. Dobbin, 'The story of the formula for pyridine', *J. Chem. Educ.* **11** (1934) 596–600, and by A.J. Rocke, 'Koerner, Dewar and the structure of pyridine', *Bull. Hist. Chem.* **2** (1988) 4–6, who cites Koerner's paper as *Giornale di scienze naturali ed economiche, Palermo* **5** (1869) 111–14. Dewar's paper is note 17, above.

26 The letter is in Kekulé's *Sammlung* at Darmstadt. It is quoted and, in part, reproduced by Rocke, note 25, above.

27 *RIA* MM, 8/6/1925.

28 The *cumul* is often thought of as a French nineteenth-century academic practice but it was prevalent also in Britain, and for the same reason, many posts were inadequately endowed to support a man and his family. Dewar was rarely without more than one post.

29 Meiklejohn, see Chapter 1, note 3.

30 J.G. McKendrick (1841–1926); *ODNB*; D.N. P[aton], *PRS B* **100** (1926) xiv–xviii; J.G. McKendrick, *The Story of My Life* (Wyllie, Aberdeen, 1919). He wrote his name indifferently with the prefix Mac, Mc and M', but the distinction is of no consequence.

31 W. Thomson, Lord Kelvin from 1892 (1824–1907); *DSB*; *ODNB*; J. L[armor], *PRS A* **81** (1908) iii–lxxvi; C. Smith and M.N. Wise, *Energy and Empire: A Biographical Study of Lord Kelvin* (CUP, 1989).

32 The title of the papers in this series by J. Dewar and J.G. McKendrick is 'The physiological action of light' but the numbering into parts confuses papers published in different journals; see *PRSE* **8** (1872–5) 100–4, 110–14, 179–82, 513, 534; *TRSE* **27** (1873) 141–66; *PRI* **7** (1875–6) 360–7, **8** (1876) 137–49, which correspond to *CP* 31–4, 44–6, 48, some by title only. They also presented their early work at the

meeting of the British Association at Bradford in 1873, *CP* 30 (title only), and the first three papers as a pamphlet printed in 1873.

33 T.H. Huxley (1825–95); *DSB*; *ODNB*; M. F[oster] *PRS* **59** (1895–6) xlvi–lxvi. For his time in Edinburgh in 1875 and 1876, see L. Huxley, *Life and Letters of Thomas Henry Huxley* (Macmillan, London, 1900) v. 1, pp. 442–3, 456–8, and A. Desmond, *Huxley: Evolution's High Priest* (Joseph, London, 1988) pp. 67, 77–8, 87.

34 J.G. McKendrick to H.E Armstrong, 21/7/1923, *RIA* DB 11/1/3.

35 Dewar's Friday Evening Discourses are published in *PRI*, note 32, *CP* 46 and 48. He had earlier given a Discourse on 7 March 1873 on 'Temperature of the Sun and work of sunlight', *PRI* **7** (1873) 57 (title only), *CP* 28.

36 J. Dewar and J.G. McKendrick, 'On the physiological action of ozone', *PRSE* **8** (1875) 211–19, *CP* 35, and as a pamphlet (Neill, Edinburgh, 1874) p. 8. Neill was the publisher to the Royal Society of Edinburgh.

37 T. Graham, 'On the relation of hydrogen to palladium' and 'Additional observations on hydrogenium', *PRS* **17** (1869) 212–20, 500–6.

38 J. Dewar, 'Motion of a palladium plate during the formation of Graham's hydrogenium', *PRSE* **6** (1869) 504–12, 'Note on the specific heat of hydrogenium', *PM* **44** (1872) 400; 'On the physical constants of hydrogenium', *TRSE* **27** (1873) 167–73, *PM* **47** (1874) 334, *CP* 5, 24 and 25. Notes of his experiments on 'hydrogenium' are in his notebook *RIA* DB I/2 pp. 53 and 61–73.

39 J. Dewar and P.G. Tait, 'Preliminary note on a new method of obtaining very perfect vacua', and 'Farther researches in very perfect vacua' *PRSE* **8** (1872–5) 348–9, 628 (Abstract only). The first paper was given before the Society in March 1874 and the second was read by Dewar in July 1875, but no full account was submitted. There is, however, a long summary by Scott-Lang, Tait's assistant, 'Charcoal vacua', in *Nature* **12** (1875) 217–18, which is reprinted as *CP* 43. Kelvin thought that this summary did not do justice to the completeness of Dewar and Tait's explanation of the operation of Crookes's radiometer, see Knott, note 3, above, quoting Kelvin's obituary of Tait in *PRSE* **23** (1899–1901) 498–502.

40 W. Crookes (1832–1919); *DSB*; *ODNB*; W.A. T[ilden] *PRS A* **96** (1919–20) i–ix; W.H. Brock, *William Crookes (1832–1919) and the Commercialization of Science* (Ashgate, Aldershot, 2008) pp. 162–78.

41 Crookes's radiometers were made for him by his assistant Charles Gimingham and were later given to the Royal Society who deposited them in the Science Museum; see J. Wess, 'Crookes's radiometers: a train of thought manifest', *NRRS* **64** (2010) 457–70.

42 W. Crookes, 'On the action of heat on gravitating masses', *PRS* **22** (1873) 37–41; 'On attraction and repulsion resulting from radiation', *PTRS* **164** (1874) 501–27; 'On the attraction and repulsion accompanying radiation', *PM* **48** (1874) 81–95.

43 J.C. Maxwell, *Treatise on Electricity and Magnetism* (OUP, 1873) v. 2, §793.

44 The discussion of Crookes's results by Reynolds and others is summarised by Brock, note 40, above.

45 Norman Lockyer (1836–1920); *DSB*; *ODNB*; A. F[owler] *PRS A* **104** (1923) i–xiv; A.J. Meadows, *Science and Controversy: A Biography of Sir Norman Lockyer* (2nd edn, Macmillan, London, 2008) p. 34.

46 Tait's letter to Maxwell of 26 June 1878 and Maxwell's reply of 8 July are printed by P.M. Harman, *The Scientific Letters and Papers of James Clerk Maxwell* (CUP, 2003) v. 3, pp. 683–6, No. 708. Tait's letter, together with much of Maxwell's

further writing on the radiometer and on stresses in rarefied gases, is also printed in E. Garber, S.G. Brush and C.W.F. Everitt, *Maxwell on Heat and Thermodynamics* (Lehigh University Press, Bethlehem, PA, 1995) Sec. VI, pp. 387–494. The éloge of Stokes by Tait appeared in the same issue of *Nature* as the account of the paper by Dewar and Tait.

47 See, for example, E.H. Kennard, *Kinetic Theory of Gases* (McGraw-Hill, New York, 1938) pp. 327–37.

48 W. Dittmar (1833–92) FRS; *DSB*; A.C[rum]B[rown], 'William Dittmar', *Nature* **45** (1892) 493–4.

49 J. Dewar and W. Dittmar, 'On the vapour-density of potassium', *PRS* **21** (1873) 203–4; *CP* 26.

50 J. Dewar and A. Scott, 'On the vapour-densities of potassium and sodium', *PRS* **29** (1879) 206–9, 490–3; *CP* 68, 72.

51 H.E. Roscoe, Presidential Address to Section B of the BAAS, *Rep. BAAS* **54** (1884) 659–69. Henry Roscoe (1833–1915); *DSB*; *ODNB*; T.E. T[horpe], *PRS A* **93** (1916–17) i–xxi. [Franz] Arthur [Friedrich] Schuster (1851–1934) a German-born spectroscopist who worked at Owens College, now Manchester University, and became a British subject in 1875. *DSB*; *ODNB*; G.C. Simpson, *ONFRS* **1** (1932–5) 251–62.

52 Thomas Anderson (1819–74); *DSB*; *ODNB*.

53 Meiklejohn, see Chapter 1, note 3.

54 Notes on some of his analyses of manures and of water are in his notebook *RIA* DB 1/2.

55 J. Dewar, 'Note on a new Scotch acidulous chalybeate mineral water', *PRSE* **7** (1872) 470–1; *CP* 16 (Chalybeates contain iron).

56 'Notes', 13/5/1875, *Nature* **12** (1875) 33.

57 'Notes', 23/4/1874, *Nature* **9** (1873–4) 490.

58 *Testimonials in favour of James Dewar ... Candidate for the Chair of Chemistry at Glasgow*, printed privately, 1874. Many of the originals are to be found with Dewar's papers at the Royal Institution.

Chapter 3 Cambridge

1 J.R. Tanner, ed., *The Historical Register of the University of Cambridge ... 1910* (CUP, 1917); *CUR* 39.20/1.

2 R. Willis (1800–75); *ODNB*.

3 The legislation and government of the University was in the hands of three bodies: Senate – the legislative body of all Masters of Arts, resident or not; Regent House – all teaching MAs in the University and some senior administrative staff, approximately 300 in number; Council of Senate – the Vice-Chancellor, four heads of colleges, four professors, and eight others, all by election. This was the executive body. (The corresponding bodies at Oxford were the Houses of Convocation and Congregation, and the Hebdomadal Council.)

4 *CUR* 39.20/11; CU *Reporter*, 9/3/1875.

5 'Notes', 11/3/1875, *Nature* **11** (1874–5) 373.

6 A. Freeman, *CUR* 39.20/10.

7 G.D. Liveing (1827–1924); *ODNB*: C.T. Heycock, *JCS Trans* **127** (1925) 2982–4; C.T. H[eycock], *PRS* **109** (1925) xxviii–xxix; J. Shorter, 'Chemistry at Cambridge under George Downing Liveing', chap. 7 of M.D. Archer and C.D. Haley, eds, *The 1702 Chair of Chemistry at Cambridge: Transformation and Change* (CUP, 2005).

8 G.M. Humphry (1820–96); *ODNB*; T.C. A[llbutt], *PRS* **75** (1905) 128–30.

9 G.G. Stokes (1819–1903); *DSB*; *ODNB*; [Lord] R[ayleigh], PRS **75** (1905) 199–216; J. Larmor, *Memoir and Scientific Correspondence' of Sir George Gabriel Stokes* (CUP, 1907) 2 v.

10 Tait to Maxwell, 26/6/1878, see Chapter 2, note 46. Tait, a nationalistic Scot, may have disapproved of Stokes's attempt to lure Dewar to England.

11 *CUR* 39.20/12; CU *Reporter*, 16/3/1875.

12 *CUR* 39.20/18.

13 This seems also to have been a fashionable view in the 1870s; see eg. G.D. Liveing, 'Natural Sciences Tripos', *Students' Guide to the University of Cambridge* (3rd edn, Deighton Bell, Cambridge, 1874) pp. 218–36, and A. Crum Brown, Presidential Address to the Chemistry Section, *Rep. BAAS Trans. Sec.* **44** (1874) 45–50. Perhaps a more modern choice of words would be Maxwell's assertion the previous year that chemistry was a 'physical science'; see his manuscript 'Remarks on the classification of the physical sciences' printed by Harman, see Chapter 2, note 46, no. 377, v. 2, pp. 654–60, and repeated in Maxwell's posthumous article on 'Physical sciences' in the 9th edition of *Encyclopaedia Britannica* (1885). Dewar's later career was to be at the conjunction of the two subjects of physics and chemistry.

14 J.W. Strutt (1842–1919), Lord Rayleigh from 1873, A. Schuster, *PRS A* **98** (1920–1) i–l; R.J. Strutt [4th Lord Rayleigh], *Life of John William Strutt, 3rd Baron Rayleigh, O.M., F.R.S.* (original edition, 1924, augmented edition, University of Wisconsin Press, Madison, WI, 1968).

15 C.T. Heycock to H.E. Armstrong, 3/2/1924, *RIA* DB 11/1/18.

16 Dewar to Stokes, 6/4/1875, CUL Add. Ms. 7342, D.B. Wilson, *Catalogue of the Manuscript Collections of Sir George Gabriel Stokes and Sir William Thomson, Baron Kelvin of Largs in the Cambridge University Library* (CUP, 1976) D 432.

17 James Douglas Hamilton Dickson (1849–1931); *ODNB*.

18 J.D. Hamilton Dickson, 'James Dewar, 1842–1923', *Peterhouse Magazine*, Easter Term, 1923, Cambridge; *RIA* D 11/1/12a.

19 *CUR* 39.20/28 and 29; *The Times*, 8/4/1875.

20 H.E. Armstrong; *DSB*; *ODNB*; F.W. Keeble, *ONFRS* **3** (1939–41) 229–45; H. Hartley, 'First Armstrong Memorial Lecture of the Society of Chemical Industry', delivered at the Royal Institution, 21/11/1945, and printed as chap. 9 of his *Studies in the History of Chemistry* (OUP, 1971); J.V. Eyre, *Henry Armstrong, 1848–1937* (Butterworth, London, 1958), esp. p. 59.

21 J.C.W. Ellis, Sydney Sussex College, bracketed 3rd Wrangler in 1856.

22 W.N. Hartley; *ODNB*; J.Y. B[uchanan], *PRS A* **90** (1914) vi–xiii.

23 E.J. Mills; see *Nature* **107** (1921) 432–3. He had been elected to the Royal Society the year before the Jacksonian chair became vacant but he never received an obituary.

24 J. Stuart; *ODNB*.

25 Liveing to Armstrong, 26/3/1875, Armstrong papers, Imperial College, II: C.414, quoted by G.K. Roberts, 'The liberally educated chemist: Chemistry at Cambridge in the Natural Sciences Tripos', *Hist. Stud. Phys. Biol. Sci.* **11** (1980) 157–83.

26 *CUA*; Elections to the Jacksonian Chair O.XIV.53. The title of the chair in successive editions of the *Calendar* was Professor of Natural Experimental Philosophy, but later it was sometimes described as of Natural and Experimental Philosophy, as in the annual lists published in *Chemical News*, and informally the title was often abbreviated to Professor of Natural Philosophy or Professor of Experimental Philosophy. Perhaps more remarkable is that Dewar described himself in his will of 8 July 1922, and again in a codicil to it of 18 December 1922, as 'Jacksonian Professor of Chemistry'.

27 'Notes', 15/4/1875, *Nature* **11** (1874–5) 474.

28 Playfair to Dewar, 14/4/1875, *RIA* D I f/10.

29 CU *Reporter*, 23/11/1875.

30 G.W. Liveing, 'Appreciation' in J. Larmor, note 9, above, pp. 91–7. The provision of chemistry laboratories in Cambridge is described by R. Willis (the late Jacksonian Professor) and J. Willis Clark (his nephew) in *The Architectural History of the University of Cambridge* (CUP, 1886), 'History of the museums and lecture-rooms for natural science', v. 3, pp. 145–90, and in the chapters of W.H. Brock, 'Coming and going: the fitful career of James Cumming', chap. 6, and J. Shorter, chap. 7, in Archer and Haley, note 7, above.

31 The name of Gonville and Caius College was and is commonly abbreviated to Caius, a convention followed here.

32 CU *Reporter*, 21/10/1879.

33 The Brookside address is on the cover of a lecture note book dating from 1875, *RIA* DE 10/4; in the list of members of the BAAS his address is given as Scroope Terrace from 1885, and the sale of this house is mentioned in a letter from Hamilton Dickson to William J. Green, Dewar's last research assistant, 9/4/1924, *RIA* DE 9/22/4.

34 Maxwell's wife, Katherine, was a Dewar, the daughter of the Principal of Marischal College, Aberdeen, but apparently no relation of James.

35 Joseph John Thomson (1856–1940); *DSB*; *ODNB*; [4th] Lord Rayleigh, *ONFRS* **3** (1939–41) 587–609; *The Life of Sir J.J. Thomson* (CUP, 1942).

36 The titles and times of his and other professors' lectures are listed at the start of each term in the Cambridge *Reporter*.

37 *Jubilee of the Chemical Society* (CS, London, 1896); *The Times* reporter described his lecture to the BAAS in Dublin on 'Dissociation; or modern ideas on chemical action' as having 'many attractive experiments' which brightened up the Chemistry Section, 'perhaps the driest in the Association'; *The Times*, 23/8/1878, while in *Nature* the lecture was praised for its 'magnificient display of experimental illustrations ... visible to an immense audience'. *Nature* **18** (1878) 437; *RIA* MM 7/4/1879. The Chemical Society lectures, 'On Dissociation', were reported in *CN* **29** (1874) 141–3, and **45** (1882) 204–6, *CP* 37 and 93. The Discourse of 21 April 1882, 'Researches of Henri Ste. Claire Deville', was not printed, *PRI* **10** (1882) 87 (title only), *CP* 105.

38 *Cambridge Calendar* and *CN* **32** (1875) 120–1.

39 *The Times*, 11/4/1876 and 5/5/1880.

40 Frederick Guthrie (1833–86); *ODNB*; G.C. Foster, *Nature* **35** (1886–7) 8–10; *Proc. Phys. Soc.* **8** (1886–7) 9–13. W.H. Brock, 'The chemical origins of practical physics', *Bull. Hist. Chem.* **21** (1998) 1–11.

41 *RSA* EC/1877/04.

Chapter 4 Demonstrators

1 Registrary files, CUR 39.20/35. Most of the documents involved in the subsequent arguments over the dismissal of Siegfried Ruhemann, Dewar's second assistant, are printed in an untitled pamphlet issued by Dewar on 5 October 1891. Copies of this are with Dewar's papers at the Royal Institution (DC 2/83) and in Cambridge University Library. It is cited here as Dewar, *Pamphlet*.
2 Alexander Scott (1852–1947); R. Robertson, *ONFRS* **6** (1948–9) 251–62, and, in a shortened form, *JCS* (1950) 762–7.
3 *RSA* EC 1898/14.
4 A. Scott, 1902, at a lecture to the Chemical Club at Cambridge; A.J. Berry and E.A. Moelwyn-Hughes, *PCS* (1963) 357–63.
5 August Wilhelm von Hofmann (1818–92); *DSB*; *ODNB*.
6 Siegfried Ruhemann; R.S. Morrell, 'Siegfried Ruhemann. 1859–1943', *JCS* (1944) 46–8; [A.]B. Pippard, 'Siegfried Ruhemann (1859–1943), FRS 1914-1923', *NRRS* **47** (1993) 271–6; M.D. Saltzmann, 'Is science a brotherhood? The case of Siegfried Ruhemann', *Bull. Hist. Chem.* **25** (2000) 116–21. The Registrary Files, *CUR* 39.20/38, report the appointment of 'S. Richemann', but this spelling error probably arose from Dewar's bad handwriting.
7 *CUR* 39.20/40.
8 J.W. Strutt (1842–1919), 3rd Lord Rayleigh from 1873; *DSB*; *ODNB*; A. Schuster, *PRS A* **98** (1920–1) i–l; R.J. Strutt [4th Lord Rayleigh], see Chapter 3, note 14, pp. 103, 412.
9 W.H. Brock, see Chapter 2, note 40, p. 296.
10 *RIA* DC 2/81. Matthew Moncrieff Pattison Muir (1848–1931) was a Glasgow graduate who had taught at Anderson's College and at Owens College, Manchester, before coming to Cambridge in 1877, where he was a Fellow of Caius and in charge of their chemical laboratory until his retirement in 1908.
11 *RIA* DC 2/21.
12 *CUR* 39.20/50; John Ebenezer Foster (1847–1912), Trinity, an antiquary and local solicitor.
13 Dewar to Thomson, 15/10/1891. D.B. Wilson, *Catalogue of the Manuscript Collections of Sir George Gabriel Stokes and Sir William Thomson, Baron Kelvin of Largs in the Cambridge University Library* (CUP, 1976) Add. Ms. 7342, Letter D 82.
14 Michael Foster (1836–1907); *DSB*; *ODNB*. W.H. G[askell], *PRS B* **80** (1908) lxxi–lxxxi.
15 M. Ruhemann, *The Separation of Gases* (OUP, 1940). Martin Ruhemann wrote but never published an account of his life before the outbreak of the Second World War, when he was 36. In it he says little about his father's row with Dewar but makes it clear that his father did not have a high opinion of the cultural life of Cambridge in general and of Peterhouse in particular. I thank Kay Baxandall for the loan of Ruhemann's manuscript.
16 *CUR* 39.20/45.
17 *CUR* 39.20/46, CU *Reporter*, 26/1/1892.
18 *RIA* DC 2/22 and 23.
19 *CUR* 39.20/54 and *RIA* DC 1/1/2.
20 *RIA* DC 2/23a.

21 *Ordinances of the University of Cambridge* (CUP, 1923).
22 William Thomas Newton Spivey (1868–1901), T.B. W[ood], *JCS* (1902) 535–6.
23 H.O. Jones; J. Shorter, 'Humphrey Owen Jones, F.R.S. (1878–1912), chemist and mountaineer', *NRRS* **33** (1979) 261–77. Henry John Horstman Fenton (1854–1929) *ODNB*; Demonstrator (1878–1904) and University Lecturer (1904–24) in Chemistry, Cambridge; W.H. M[ills], *PRS A* **127** (1930) i–v.
24 *RIA* D II a/48 and 49.
25 *RS* Minutes of Council, 3/10/1912.
26 William Hodson Mills (1873–1959), F.G. Mann, *BMFRS* **6** (1960) 201–25.

Chapter 5 Spectroscopy

1 Heycock, see Chapter 3, note 7.
2 *CN* **30** (1874) 131; **32** (1875) 120–1.
3 Private letter by Ramsay, cited by M.W. Travers, *A Life of Sir William Ramsay* (Arnold, London, 1956) p. 70.
4 H.E. Armstrong, 'James Dewar', *JCS* **131** (1928) 1066–76.
5 W. Crookes, 'On the existence of a new element, probably of the sulphur group', *CN* **3** (1861) 193–4, 'Further remarks …', p. 303; C.A. Lamy, 'De l'existence d'un nouveau metal, le thallium', *CR* **54** (1862) 1255–8; F.A.J.L. James, 'Of "Medals and Muddles". The context of the discovery of thallium: William Crookes's early spectro-chemical work', *NRRS* **39** (1984) 65–90; chap. 10 of Brock, Chapter 2, note 40.
6 F. Reich and T. Richter, 'Ueber das Indium', *J. Prakt. Chem.* **90** (1863) 172–6; **92** (1864) 480–5.
7 H.E. Roscoe, 'On Bunsen and Kirchhoff's spectral observations' *PRI* **3** (1858–62) 323–8; *CN* **3** (1861) 153–5, 170–2; **5** (1862) 218–22, 261–5, 287–93.
8 W.A. Miller (1817–70) of King's College, London; *DSB*; *ODNB*; C[harles] T[ownsend], *PRS* **19** (1870–1) xix–xxvi. W.A. Miller, 'On spectrum analysis', *CN* **5** (1862) 201–13, 214–18. W. Huggins (1824–1910); *DSB*; *ODNB*; F.W. D[yson], *PRS A* **86** (1911–12) i–xix.
9 W. Huggins and W.A. Miller, 'On the spectra of some of the fixed stars', *PTRS* **154** (1864) 413–35.
10 W.J. Green, 'John William Edward Heath 1856–1934', *JCS* (1934) 2014–16. Heath worked in the RI from his appointment in 1881 until 1925, after Dewar's death in 1923.
11 Liveing to Dewar, 20–29/12/1898, 9/2/1904, 2/1/1905, 4/1/1905, 11/12/1906, 11/8/1908 in *RIA* D I b/14–21, 38–45.
12 *RIA* MM 4/3/1878. Siemens was also concerned with the constitution of the Sun, although his speculations 'On the conservation of solar energy', *PRS* **33** (1882) 389–98, were soon shown to be wide of the mark. This paper was read at a meeting of the Royal Society attended by the Prince of Wales so that he could sign the Charter Book, nearly 20 years after his election. See W. White, *The Journals of Walter White* (Chapman & Hall, London, 1898), entry for 2/3/1882, p. 276. For the electricity supplies at the RI, see D.C. Chilton and N.G. Coley, 'The laboratories of the Royal Institution in the nineteenth century', *Ambix* **27** (1980) 173–203.

13 *RIA* MM 3/2/1879. This machine was invented by Baron Auguste de Meritens (1834–98) in the 1870s and used by him to develop the technique of arc-welding in 1881.

14 J. Dewar, 'Note on electrolytic experiments', *PRS* **30** (1879–80) 170–2, *CP* 79.

15 G.W. Liveing, 'On a new spectroscope', *PCPS* **3** (1879) 260–1; 'On a spectrometer and universal goniometer adapted to the ordinary wants of the laboratory', *PCPS* **4** (1883) 343–4, *CPS* 13 and 45.

16 John Edward Purvis (1862–1930), educated at Owens College, Manchester, and Trinity College, Dublin, where Hartley was then the Professor. After Liveing's retirement he turned to preventive medicine and was later Mayor of Cambridge.

17 Tanner, see Chapter 3, note 1, pp. 241–2.

18 *RIA* MM, 5/12/1898.

19 J. Dewar, 'On recent estimates of solar temperature', *Rep. BAAS , Trans. Sec.* **42** (1872) 50–1, *PRSE* **7** (1872) 697–8; *CP* 20.

20 J. Dewar, Chemical Society Lecture, 'On dissociation', *CN* **29** (1874) 141–3; *CP* 37.

21 See, for example, John F.W. Herschel, *Preliminary Discourse on the Study of Natural Philosophy* (Longman, London, 1830) §199, 252, and the popular *The Unseen Universe* by Balfour Stewart and Guthrie Tait (Macmillan, London, 1875) who refer their readers to W.R. Grove, *The Correlation of Physical Forces* (Highley, London, 1849). H. Sainte-Claire Deville et al., 'Rapport sur un mémoire de MM Troost et Hautefeuille sur les transformations isomériques et allotropiques', *CR* **76** (1873) 1175–82, and his earlier book of 1860, see Chapter 3. By the 1870s chemists and physicists had the example of the work of Thomas Andrews and J.D. van der Waals; T. Andrews, 'On the continuity of the gaseous and liquid states of matter', *PTRS* **159** (1869) 575–90; J.D. van der Waals, *Over de Continuiteit van den Gas-en Vloeistofstand* (Thesis, Leiden, 1873), English trans. *On the Continuity of the Gaseous and Liquid States*, Studies in Statistical Mechanics, v. 14 (North-Holland, Amsterdam, 1988). Outside the physical sciences there was the powerful influence of Darwin's *Origin of Species* from 1859, which later in the century gave rise to speculation that there might also have been an 'inorganic evolution' of the elements, an idea that Crookes did much to foster, as, for example, in his Presidential Address to Section B of the British Association in Birmingham in 1886; *Rep. BAAS* **56** (1886) 558–76.

22 W.V. Farrar, 'Nineteenth-century speculation on the complexity of the chemical elements', *BJHS* **2** (1965) 297–323; W. McGucken, *Nineteenth-Century Spectroscopy: Development of the Understanding of Spectra, 1802–1897* (Johns Hopkins University Press, Baltimore, MD, 1969).

23 William Prout (1785–1850); *DSB*; *ODNB*; W.H. Brock, *From Protyle to Proton: William Prout and the Nature of Matter, 1785–1985* (Hilger, Bristol, 1985) esp. chap. 8. The name 'atomic weights' was almost universally used in the nineteenth and early twentieth centuries for what are now more properly called 'relative atomic masses'.

24 J.-S. Stas (1813–91); *DSB*: J.W. Mallet, *Memorial Lectures delivered before the Chemical Society*; *JCS Trans* (1893) 1–56, reprinted in *Memorial Lectures of the Chemical Society, 1893–1900* (Chemical Society, London, 1901). This lecture, 'Jean-Servais Stas and the measurement of the relative masses of the atoms of the chemical elements', is more than a memorial to Stas but reflects also Mallet's views on the determination of atomic masses and on the status of Prout's hypothesis.

Mallet was not present so his paper was read by Armstrong, one of the Secretaries. John William Mallet (1832–1912); *DSB*. Elected FRS in 1877, but no obituary.

25 J.W. Mallet, 'Revision of the atomic weight of aluminium', *PRS* **30** (1897–80) 329–32.

26 R.J. Strutt, 'On the tendency of the atomic weights to approximate whole numbers', *PM* **1** (1901) 311–14. Robert ['Robin'] John Strutt (1875–1947), 4th Lord Rayleigh from 1919, *ODNB*; A.C. Egerton, *ONRS* **6** (1948–9) 503–48.

27 Lord Rayleigh, President's Address to Section A of the British Association, *Rep. BAAS* **52** (1882) 437–41.

28 T.E. Thorpe, 'On the atomic weight of titanium', *PCS* **1** (1885) 1–3; Thomas Edward Thorpe (1845–1925); *DSB*; *ODNB*; A.E.H. T[utton] *PRS A* **109** (1925) xviii–xxiv.

29 He had spoken on this subject two years previously in his President's Address to Section B of the British Association, *Rep. BAAS* **53** (1883) 448–54.

30 W. Crookes, President's Address to Section B of the British Association, *Rep. BAAS* **56** (1886) 115–26; President's Address to the Chemical Society, 'On elements and meta-elements', *JCS* **53** (1888) 487–504.

31 Charles Augustus Young (1834–1908); *DSB*. C.A. Young, 'Catalogue of the bright lines in the spectra of the Solar atmosphere', *Nature* **7** (1872) 17–20; 'Spectroscopic nomenclature', **7** (1872) 101; 'Spectroscopic notes, 1879–1880', *(Silliman's) Amer. J. Sci.* **20** (1880) 353–8.

32 J.N. Lockyer, Bakerian Lecture, 27 November 1873, 'Researches in spectrum analysis in connexion with the spectrum of the Sun, No. III', *PTRS* **164** (1874) 479–94; *Studies in Spectrum Analysis* (Kegan Paul, London, 1878); The development of his ideas from 1872 onwards is described in chapter 7 of Lockyer's book, and by H. Dingle in chapter 10 of T.M. and W.L. Lockyer, *Life and Work of Sir Norman Lockyer* (Macmillan, London, 1928); W.H. Brock, 'Lockyer and the chemists: the first dissociation hypothesis', *Ambix* **16** (1969) 81–99. At the time of his Bakerian Lecture Lockyer was busy as Secretary of the Royal Commission on Scientific Instruction and the Advancement of Science, Eight Reports, 1871–5, commonly known as the reports of the Devonshire Commission. The question has therefore been raised about the extent to which the ideas expressed there were those of his assistant R.J. Friswell, whose help he acknowledges but whose collaboration receives no greater mention, as was the custom of the time for paid assistants. For Richard John Friswell (1849–1908), see 'Notes', *Nature* **77** (1908) 349 and A.J.G., *JCS Trans.* **95** (1909) 2204–6.

33 J.N. Lockyer, 'Researches in spectrum analysis in connexion with the spectrum of the Sun, No. VII', *PRS* **28** (1878–9) 157–80; reprinted by Crookes under the title of the opening section as 'Discussion of the working hypothesis that the so-called elements are compound bodies', *CN* **39** (1879) 1–5, 11–16. A fair and anonymous summary in *The Times* was reprinted in *Nature* **19** (1878–9) 157–8.

34 Crookes to Lockyer, 16 December 1878, cited by Meadows, see Chapter 2, note 45, p. 149.

35 For Roscoe's views see his Presidential Address to Section B of the BAAS meeting in Montreal, *Rep. BAAS* **54** (1884) 659–69, and his Presidential Address at the Manchester meeting, *Rep. BAAS* **57** (1887) 1–28.

36 Stokes to Lockyer, 27 January 1879, reprinted by J. Larmor, see Chapter 3, note 9, v. 1, pp. 398–432, see 406–7.

37 J.N. Lockyer, 'Note on the existence of carbon in the coronal atmosphere of the Sun', *PRS* **27** (1878) 308–9.

38 G.D. Liveing and J. Dewar, 'On the reversal of the lines of metallic vapours, No. V', *PRS* **28** (1878–9) 367–72, *CPS* 8; J.N. Lockyer, 'Note on a recent communication of Messrs. Liveing and Dewar', *PRS* **29** (1879) 45–7; G.D. Liveing and J. Dewar, 'Note on "Spectroscopic papers"', *PRS* **29** (1879) 166–8, *CPS* 14.

39 Meadows, see Chapter 2, note 45, p. 149–55. William Chandler Roberts [later Roberts-Austen] (1843–1902); *DSB*; *ODNB*; T.E. T[horpe], *PRS* **75** (1905) 192–8.

40 J.N. Lockyer, 'Notes on recent spectral observations', Abstract in *Rep. BAAS* **49** (1879) 317–18; the paper appeared as, 'On a new method of studying metallic vapours', *PRS* **29** (1879) 266–72; submitted 19 June 1879.

41 Roscoe to Lockyer, 11 April 1880, cited by Meadows, see Chapter 2, note 45, p. 152.

42 C.A. Young, 'Spectroscopic Notes, 1879–1880', (*Silliman's*) *Amer. J. Sci.* **20** (1880) 353–8.

43 G.D. Liveing and J. Dewar, 'On the identity of spectral lines of different elements', *PRS* **32** (1881) 225–31, read 12 May 1881, *CPS* 25. Dewar returned to this subject in a Friday Discourse at the RI, 'On the origin and identity of spectra', *PRI* **9** (1881) 674–703, *CPS* 27. This discourse contains the typical phrase: 'Now what are the facts?', which he had also recently used in a paper with Liveing: 'Investigations on the spectrum of magnesium', *PRS* **32** (1881) 189–203, and which also occurs elsewhere.

44 The discussion was organised by Crum Brown and held in 1885 at the BAAS meeting at Aberdeen; see *The Times*, 21/9/1885, and *Nature* **32** (1885) 533–4.

45 The earliest of Stokes's letters of advice to Dewar on spectroscopy seems to be that written from Brookside Terrace in Cambridge on 11 February 1878; D.B. Wilson, ed., *Catalogue of the Manuscript Collection of Sir George Gabriel Stokes and Sir William Thomson, Baron Kelvin of Largs, in the Cambridge University Library* (CUP, 1976) Letter D 433. This letter was written a week before the first joint paper of Liveing and Dewar was read to the Royal Society. McKendrick said in a letter to Armstrong, 21 July 1923, that Stokes was one of the few men for whom Dewar had 'an infinite regard', *RIA* DB 11/1/3.

46 See *Rep. BAAS* **50** (1880) 499–500.

47 See *Rep. BAAS* **51** (1881) 317–423; **54** (1884) 295–350, 351–446.

48 William Abney (1843–1920) was an officer in the Royal Engineers, an authority on photography and astronomy who was commonly styled Captain Abney; *DSB*; *ODNB*; E.H. G[rove]-H[ills], *PRS A* **99** (1921) i–v. For Arthur Schuster, see Chapter 2, note 51.

49 Dewar's Presidential Address at Sheffield is listed by title in *Rep. BAAS* **49** (1879) 293, but there is an anonymous account in *CP* 69. For Liveing at Southampton, see *The Times*, 18/8/1882.

50 Dewar's paper at York, 'On the alleged decomposition of the elements' is listed by title in *Rep. BAAS* **51** (1881) 589, *CP* 89, but there is an account in *The Times*, 8/9/1881, and its substance is in *CPS*, pp. 156–62 and 174–5.

51 Lord Rayleigh, President's Address, *Rep. BAAS* **54** (1884) 1–23; Strutt, *Rayleigh*, see Chapter 3, note 14, p. 165.

52 Josiah Willard Gibbs (1839–1903); *DSB*; J.L[armor] *PRS* **75** (1905) 280–96. Letters and reprint lists in L.P. Wheeler, *Josiah Willard Gibbs: The History of a*

Great Mind (Yale University Press, 1962) pp. 221, 223, 242, 261; letters and notes in P.M. Harman, ed., *The Scientific Letters and Papers of James Clerk Maxwell* (CUP, 2002) v. 3, pp. 147–8, 230–2, 236–8.

53 'The American Association', *Nature* **30** (1884) 596–602.

54 These became widely available only in 1904: Lord Kelvin, *Baltimore Lectures on Molecular Dynamics, and the Wave Theory of Light* (Clay, London, 1904); R. Kargon and P. Achinstein, *Kelvin's Baltimore Lectures and Modern Theoretical Physics* (MIT Press, Cambridge, MA, 1987).

55 His American travels led to a long and erudite lecture to the Society of Arts the next year; J. Dewar, 'The American oil and gas fields', *J. Soc. Arts* **33** (1885) 771–80; *CP* 124.

56 G.D. Liveing and J. Dewar, 'On the ultra-violet spectra of the elements', *PTRS* **174** (1883) 187–212, 213–22; **179** (1889) 231–56, *CPS* 33 and 54.

57 A. Schuster, '"The genesis of spectra", a Report of the Committee appointed for the purpose of reporting on the present state of our knowledge of spectrum analysis', *Rep. BAAS* **52** (1882) 120–43.

58 J.J. Balmer, 'Notiz über die Spektrollinien des Wasserstoffs', *Ann. Physik* **25** (1885) 80–7.

59 J.R. Rydberg, 'Sur la constitution des spectres liniéaires des éléments chimiques', *CR* **110** (1890) 394–7.

60 For 'sharp' and 'diffuse', see many of their papers, starting with *CPS* 17; for 'principal', see e.g. *CPS* 14 and 41.

61 A. Bergmann, *Beiträge zur Kenntnis der ultrarothen Emmissionspectren der Alcalien*, Thesis, Jena, cited by W.B. Jensen, 'The origin of the s, p, d, f orbital labels', *J. Chem. Educ.* **84** (2007) 757–8.

62 I thank my colleague John Eland for a discussion on the choice of the label 'fundamental'.

Chapter 6 London

1 *The Charter Acts of Parliament and Endowments of the Royal Institution of Great Britain and Supplemental Charter* (Royal Institution, London, 1967); H.B. Jones, *The Royal Institution: its Founders and its First Professors* (Longmans, London, 1871); [H.E. Armstrong] 'Chemical Laboratories of the Royal Institution', *Nature* **66** (1902) 460–2; M. Berman, *Social Change and Scientific Organization: The Royal Institution, 1799–1844* (Heinemann, London, 1978); G. Caroe, *The Royal Institution: an Informal History* (Murray, London, 1985); *James*.

2 John Tyndall (1820–1893); *DSB*; *ODNB*; E.F[rankland], *PRS* **55** (1894) xviii–xxxiv; A.S. Eve and C.H. Creasey, *Life and Work of John Tyndall* (Macmillan, London, 1945); U. DeYoung, *A Vision of Modern Science: John Tyndall and the Role of the Scientist in Victorian Culture* (Palgrave Macmillan, New York, 2011).

3 Edward Frankland (1825–99); *DSB*; *ODNB*; C.A. Russell, *Edward Frankland: Chemistry and Controversy in Victorian England* (CUP, 1996). Although Foreign Secretary of the Royal Society from 1895 until his death, the Society never published an obituary.

4 William Odling (1829–1921); *DSB*; *ODNB*; H.B.D[ixon], *PRS A* **100** (1921–2) i–vii.

5 W. White, *The Journals of Walter White* (Chapman and Hall, London, 1898), entry for 28 September 1872, p. 255. Heinrich Debus (1824–1915), FRS, *ODNB*, was then a Lecturer at Guy's Hospital, and later Professor of Chemistry at the Royal Naval College, Greenwich.

6 John Hall Gladstone (1827–1902) *ODNB*; T.E.T[horpe], *PRS* **75** (1905) 188–92. For an account of the actions of the Managers at this time, see D. Chilton in his 'Introduction' to the facsimile publication of Volume 13 of the Managers' Minutes, see 'Abbreviations'.

7 *RIA* MM, 6 March 1876.

8 *RIA* MM, 5 February and 5 March 1877.

9 George Busk (1807–86) Naval Surgeon and zoologist; *ODNB*. Joseph Dalton Hooker (1817–1911) Botanist and explorer; *DSB*; *ODNB*; W.T. T[histelton]-D[yer], *PRS B* **85** (1912) i–xxxv; W.B. Turill, *Joseph Dalton Hooker: Botanist, Explorer and Administrator* (Thomas Nelson, London, 1963). [William] Frederick Pollock (1815–88) Barrister; *ODNB*. [Karl] Wilhelm, or [Charles] William Siemens (1823–83); *DSB*; *ODNB*; [Anon.], *PRS* **37**(1884) i–ix. He became a British subject in 1859 and in 1861 invented the regenerative gas furnace, which made his fortune, and was to become relevant to the problems of gas liquefaction. He was the younger brother of Ernst Werner Siemens, the better-known electrical engineer; the firm of Siemens Brothers was founded in 1858. William Spottiswoode (1825–83) mathematician, physicist and a director of the firm of printers, Eyre & Spottiswoode; President of the Royal Society, 1878–83; *ODNB*; A.B. K[empe], *PRS* **38** (1884–5) xxiv–xxxix.

10 R.M. MacLeod, 'The X-Club, a social network of science in late-Victorian England', *NRRS* **24** (1970) 305–22; J.V. Jensen, 'The X-Club, Fraternity of Victorian science', *BJHS* **5** (1970) 63–72; R. Barton, '"An influential set of chaps". The X-Club and Royal Society politics', *BJHS* **23** (1990) 53–81.

11 Huxley's view of the role the Royal Society would have been at odds with the entrepreneurial spirit often expected today from academic scientists. He did not wish to see 'the noble old Society exploited by enterprising commercial gents who make their profit from the application of science'. Huxley to W.H. Flower, 7/7/1883, quoted by L. Huxley, *Life and Letters of Thomas Henry Huxley* (Macmillan, London, 1900) v. 2, p. 52.

12 *RIA* MM 2/4/1877. The lectures were given in May and June, after his appointment.

13 William Ramsay (1852–1916); *DSB*; *ODNB*; J.N. C[ollie], *PRS A* **93** (1916–17) xlii–liv; Travers, see Chapter 5, note 3. George Farrer Rodwell (1843–1905). Alfred Tribe (1840–85), J.H. Gladstone, *Nature* **33** (1885) 180. Charles Romley Alder Wright (1844–94) *ODNB*; T.E. T[horpe], *PRS* **57** (1894–5) v–vii.

14 Turill, note 9, p. 166.

15 *RIA* MM, 2/4/1877 and 9/4/1877.

16 *RIA* MM, 4/6/1877. Warren de la Rue (1815–89) FRS, chemist, astronomer and photographer; *ODNB*. He always signed himself with a small 'd' and a small 'l', but the records of the RI usually have a capital 'D' and always index him under this letter. I follow *ODNB* and index him under 'R'.

17 *RIA* MM, 4/3/1878.

18 J. Dewar, 'Experiments on electro-photometry', *PRI* **8** (1875–8) 565–7, *CP* 52.

19 B.H. Becker, *Scientific London* (Henry King, London, 1874; reprinted by Frank Cass, London, 1968) pp. 50–2.

20 *RIA* D III a/6. A full list was printed with the proceedings of a commemorative meeting at the RI after Dewar's death, *PRI* **14** (1924) 81–6, and a list of some of the discourses by the more famous of his contempories is in J.M. Thomas, 'Picking winners: W.H. and W.L. Bragg at the Royal Institution', *NRRS* **65** (2011) 163–82.

21 Tyndall to Spottiswoode, 30/3/78, *RIA* JT/1/T/1318; Tyndall's *Journal*, *RIA* MS JT/2/16, 3/1/1884. From the date of the entry this lecture must have been one of Dewar's Christmas Lectures for children on a 'Soap Bubble'.

22 To be precise, he had then first liquefied oxygen in bulk, see J. Dewar, 'Researches on liquefied gases', *PRI* **11** (1884) 148–53; 'On the liquefaction of oxygen and the critical volumes of fluids', *PM* **18** (1884) 210–16, *CP* 122–3.

23 T.A. Hirst, *Journal*, v. 16, 5/6/1885, *RIA*. Thomas Archer Hirst (1830–92) *ODNB*; geometer, Director of Studies at the Royal Naval College, Greenwich, and also a member of the X-Club; A.R. F[orsyth] and J. T[yndall], *PRS* **52** (1892–3) xii–xviii. Dewar's Discourse, 'Liquid air and the zero of absolute temperature', is listed by title only in *PRI* **11** (1885) 318, *CP* 125.

24 McKendrick to Armstrong, 21 August 1923, *RIA* DB 11/1/3.

25 White, note 5, entry for 6 December 1878, p. 270.

26 Gerrard Ansdell (1853–1936) was one of the six sons of Richard Ansdell, a well-known painter of animal studies from Liverpool. Gerrard had trained as an analytical chemist under Carl Fresenius in his semi-private laboratory at Wiesbaden, worked at St George's Hospital Laboratory, and was elected a Fellow of the Chemical Society on 17 February 1876. After he left the employ of the RI in 1881 he went into private practice as an analytical and consulting chemist in Kensington. In 1892 he sold a camera to the RI for £4. He was, for one year, 1896–7, a Visitor at the RI. Chemical Society certificate, 20/2/1876; *RIA* MM 4/11/1878, 2/6/1879, 7/6/1880, 6/12/1880, 7/11/1881, 5/12/1892, 13/4/1896; *The Times*, 19/8/1884, 4/5/1886, 23/7/1886, 12/2/1908; Census Returns 1871–1911. I thank Sarah Kellam (née Ansdell) for much information on the Ansdell family.

27 This pump was in addition to de la Rue's gift of £50 reported to the Managers on 4 March 1878; *RIA* MM 4/3/1878.

28 J.S. Rowlinson, 'The work of Thomas Andrews and James Thomson on the liquefaction of gases', *NRRS* **57** (2003) 143–59.

29 J. Dewar, 'The liquefaction of gases' *PRI* **8** (1878) 657–63, *CP* 57.

30 J. Dewar, 'On the critical points of mixed vapours', *PRS* **30** (1880) 538–46, *CP* 83.

31 G. Ansdell, 'The physical properties of liquid acetylene', *PRS* **29** (1879) 209–14; 'The physical constants of hydrochloric acid', *PRS* **30** (1879–80) 117–22; 'On the critical points of mixed gases', *PRS* **34** (1882–3) 113–19; with J. Dewar, 'On the gaseous constituents of meteorites', *PRS* **40** (1886) 549–59; and J. Dewar, 'Recent researches on meteorites', *PRI* **11** (1884–6) 541–50, *CP* 128–9.

32 *RIA* MM 1/3/1880, 5/4/1880, 9/4/1880.

33 *RIA* MM 4/6/1877, 4/3/1878, 2/12/1878; Annual Meeting, 2/5/1892, *PRI* **13** (1890–2) 600.

34 Tyndall's *Journal*, *RIA* MS JT/2/16, 2/5/1884.

35 *RIA* MM 3/4/1882, 6/11/1882 4/12/1882, 5/2/1883, 5/3/1883, 5/11/1883, 4/2/1884, 3/3/1884, 2/6/1884.

36 Tyndall's *Journal*, *RIA* MS JT/2/15, 28/12/1883.

37 Caroe, note 1, p. 129.
38 *RIA* MM 6/6/1881, 4/7/1881.
39 *RIA* MM 5/4/1886, 5/7/1886.
40 Tyndall's *Journal*, *RIA* MS JT/2/16, 13/3/1884, MM 5/4/1886; Eve and Creasey, note 2, p. 248. Tyndall to Hirst, 1/5/1888, *RIA* JT/1/T/778. Frederick Joseph Bramwell (1818–1903) FRS, *ODNB*; Secretary of the RI from 1885 to 1900, was a consulting engineer who specialised in patent cases. He did not work well with Tyndall in his last days at the RI and so became a supporter of Dewar in the eyes of many of their contemporaries.
41 Tyndall to Spencer, cited by Eve and Creasey, note 2, p. 248.
42 *RIA* MM 2/3/1887, 7/3/1887, 4/4/1887.
43 Tyndall to Hirst, 24/3/1888, *RIA* JT/1/T/766.
44 T.A. Hirst, *Journal*, v. 16, 1/6/1887, *RIA*.
45 *RIA* MM 21/3/1887.
46 *RIA* MM 21/3/1887, 4/4/1887, 9/5/1887.
47 *RIA* MM 5/12/1887; JT 2/50.
48 J. Crichton Browne and J. Dewar (independently), *PRI* **14** (1893–5) 161–8.
49 Lord Rayleigh, 'The scientific work of Tyndall', *PRI* **14** (1893–5) 216–24.
50 *RIA* MS JT/5/3.
51 R. Cory, 'Fifty years at the Royal Institution', *PRI* **34** (1950) 661–73, reprinted as Appendix 1 in *James*.
52 For Mond's gifts, see MM 4/7/1887, 2/7/1888, 3/6/1889, 7/4/1890, 6/4/1891, 7/12/1892, 3/7/1893, 5/3/1894 etc.
53 MM 5/10/1892, 10/4/1893; W.H. Brock, *The Case of the Poisonous Socks, Tales from Chemistry* (RSC, 2011) pp. 24–39; A.M. Clerke, 'Low-temperature research at the Royal Institution, 1893–1900', *PRI* **16** (1901) 699–718; H.E. Armstrong, 'Low-temperature research at the Royal Institution, 1900–1907', *PRI* **19** (1909) 354–412.
54 R.M. Macleod, 'The Royal Society and the Government Grant: notes on the administration of scientific research, 1849–1914', *Hist. J.* **14** (1971) 323–558; *Public Science and Public Policy in Victorian England* (Ashgate, Aldershot, 1996). There is no record of any further grant to Dewar in *RSA* CMB 66.
55 W. Spottiswoode, 'On the old and new laboratories at the Royal Institution', *PRI* **7** (1873) 1–11, *Nature* **7** (1873) 223–4, 263–5; D. Chilton and N.G. Coley, 'The laboratories of the Royal Institution in the nineteenth century', *Ambix* **27** (1980) 173–203.
56 *RIA* MM 26/6/1884.
57 Strutt, see Chapter 3, note 14, p. 262.
58 Brock, see Chapter 2, note 40, p. 443.

Chapter 7 Commerce

1 Dewar to de la Rue, 1880, cited in 1976 by D. Chilton in his Introduction to Volume 14 of the reprint of the minutes of the Managers' Meetings, see 'Abbreviations'.
2 *The Plating Company* v. *Farquharson (trading as the Nickel Plating Company)*; *The Times*, 18/2/1881.

3 *Badische Anilin und Soda Fabrik* v. *Levinstein, Reports of Patent, Design and Trade Mark Cases* **2** (1885) 73–118. M.R. Fox, *Dye-Makers of Great Britain, 1856–1976: A History of Chemical Companies, Products and Change* (ICI Ltd, Blackley, Manchester, 1987) pp. 14, 18–20.

4 Ludwig Mond (1839–1909) FRS, *DSB*, *ODNB*; F.G. Donnan, *Ludwig Mond, 1839–1909*, delivered as a Friday Evening Discourse at the Royal Institution, 15/3/1939, as a lecture to the Institute of Chemistry, 12/5/1939, and then published by the Institute; J.M. Cohen, *The Life of Ludwig Mond* (Methuen, London, 1956).

5 J. Dewar, pp. 35 and 36, L. Mond, p. 40, and A. Nobel, p. 194 of the *Rep. BAAS, Trans. Sec.* **38** (1868).

6 R.W. Tolf, *The Russian Rockefellers: The Saga of the Nobel Family and the Russian Oil Industry* (Hoover Inst. Press, Stanford, CA, 1976).

7 Alfred Bernhard Nobel (1833–96); *DSB*; H. Schück and R. Sohlman, *The Life of Alfred Nobel* (Heinemann, London, 1929).

8 More properly, glyceryl trinitrate, but the name nitroglycerine is commonly used. Guncotton or nitrocellulose is, similarly, a fully nitrated cellulose. Kieselguhr is a naturally occurring diatomaceous earth. Gelignite is another high explosive invented by Nobel as a safer alternative to dynamite, in which a mixture of guncotton and nitroglycerine, and sometimes potassium nitrate, is absorbed into sawdust to preserve its stability and avoid the fault of loss of nitroglycerine by 'weeping' to which dynamite was prone.

9 S. Mauskopf, *Powders, Patents and Profits* (Chicago University Press, in press) chap. 4. I am greatly indebted to Sy Mauskopf for letting me see a draft manuscript of this chapter before publication and for sending me photocopies of Dewar's and Abel's many letters to Alfred Nobel and his company.

10 F.D. Miles, *A History of Research in the Nobel Division of ICI* (ICI Ltd, Ardeer, 1955) chap. 1; W.J. Reader, *Imperial Chemical Industries: A History*, v. 1, *The Forerunners 1870–1926* (OUP, 1970) chap. 7.

11 Frederick Augustus Abel (1827–1902) *ODNB*. He has not been well-served by his obituarists. The Royal Society never published a memoir and those published in the chemical journals deal with his many honours but only trivially with his scientific work: J. Spiller, *JCS* (1905) 565–70; K.R. Webb, *Jour. Roy. Inst. Chem.* **82** (1958) 147–50. He was trained under Hofmann and from 1854 to 1888 he was Chemist of the War Department at Woolwich Arsenal and at the Waltham Abbey Gunpowder Factory. He became a member of the RI in 1884 and a Manager in 1885.

12 Robert Hanbury, in the debate on the Army Estimates in 1893, alleged that Abel had taken out a patent to cover his work on guncotton which he had, unknown to the War Office, then sold to Messrs Prentice of Stowmarket: *Hansard*, House of Commons, 11/9/1893.

13 *RIA* D II d/2; D V b/1; DB 4/2. It seems that Dewar's assistant, Gerrard Ansdell made some of these analyses. A distillation of many gallons of Black Sea Oil is recorded in the notebook DB 5/2. The record is dated 25–26 March, but with no year, and is not in Dewar's hand. A page on the financial affairs of the Baku enterprise is at the end of the laboratory notebook DB I/1, p. 326.

14 Letter from George M'Roberts, Works Manager at Nobel's factory at Ardeer, to Alfred Nobel, 8/11/1884, cited by Mauskopf, note 9.

15 Andrew Noble (1831–1915) *ODNB*, was a Scots physicist who worked with Abel on gunpowder. J. Henry Gilbert (1817–1901); *DSB* (see under J.B. Lawes); *ODNB*,

the chief chemist at the Rothamsted Agricultural Station, who, like Dewar, travelled to the 1884 meeting of the BAAS in Montreal; William George Armstrong (1810–1900) *ODNB*, engineer and manufacturer of armaments; for Frederick Bramwell, see Chapter 6, note 40. J. Henry Lefroy (1817–90) FRS, was Director of Ordnance from 1868 to 1870, and responsible for mapping the Earth's magnetic field in Canada.

16 August Dupré (1835–1907) *ODNB*; [H. Wilson Hake] *PRS A* **80** (1907–8) xiv–xviii; P. Dupré, 'Thudichum and Dupré: brothers-in-law', *J. Roy. Soc. Med.* **86** (1993) 417–20. Dupré was a German-born-and-trained analytical chemist, who became a British subject in 1866, and who taught at the Westminster Hospital Medical School; MRI 1900. He specialised in the analysis of water and wine and in work on explosives. A founder member of the Society of Public Analysts. He, Abel and Odling had earlier sat on a committee that looked into the safety of gelignite.

17 Schück and Sohlman, note 7, p. 55.

18 There is a full discussion of the early attempts to 'fix' nitrogen, and of the patents involved, in the paper of A.A. Breneman, 'The fixation of atmospheric nitrogen', *J. Amer. Chem. Soc.* **11** (1889) 2–27, 31–46.

19 The leading contender was probably the French 'Poudre B', which was based on picric acid, or TNP, 2,4,6 trinitrophenol; F. Abel, 'Smokeless explosives', *PRI* **13** (1890–2) 7–23, a Friday Evening Discourse at the RI on 31/1/1890; S.H. Mauskopf, 'Calorimeters and crushers: the development of instruments for measuring the behavior of military powders', in S.A. Watson, ed., *Instruments of War: Science, Research and Instruments, between Knowledge and the World* (Brill, Netherlands, 2006).

20 Jocelyn Home Thomson (1859–1908) *ODNB*; son of the Archbishop of York, ended his career as Chief Inspector of Explosives, but had a breakdown in 1907 and committed suicide the next year.

21 *RSC*, Abel Papers, HC 5001/25.

22 *Proceedings and Reports of the Explosives Committee, 1888–1891*, War Office Papers, SUPP 6/512.

23 *Final Report of the President of the Explosives Committee, 1888–1891*, War Office Papers, SUPP 5/794.

24 A.V. Newton, on behalf of Alfred Nobel, British Patent No. 1471, 1888, *Improvements in the Manufacture of Explosives*; Provisional specification, 31/1/1888, Complete specification, 24/11/1888, with later supplementary patents; 'Descriptive list of Alfred Nobel's English patents', Schück and Sohlman, note 7, pp. 307–26.

25 J.H. Thomson, Appendix VIII, *Final Report*, note 23.

26 F.A. Abel and J. Dewar, British Patents No. 5614, 1889, *An Improvement in the Manufacture of Explosives for Ammunition*; Provisional specification, 2/4/1889, Complete specification, 5/7/1890: No. 8718, *Manufacture of Explosives*; Provisional, 27/5/1889, Complete, 27/2/1890: No. 11664, *Improvements in the Manufacture of Explosives*; Provisional, 22/7/1889, Complete, 22/4/1890. The long gaps between the Provisional and Complete applications arose, in part, from an initial decision by the War Office, soon abandoned, to try to preserve the secrecy of the applications.

27 Vaseline is a non-volatile petroleum jelly devised by the British chemist Robert Cheeseborough and made by him from 1872 at a plant near New York. The company is now part of the Anglo-Dutch conglomerate, Unilever.

28 Reader, note 10.
29 *CN* **76** (1897) 274–6. For Richard Friswell, Lockyer's research assistant, see Chapter 5, note 32.
30 *RSC* Abel papers HC 5001/30 (16–18).
31 William Anderson (1834/5–1898) FRS, *ODNB*; a mechanical engineer, born in St Petersburg, trained under William Fairbairn in Manchester, and later a partner in Eaton and Anderson at Erith, on the south bank of the Thames and conveniently close to Woolwich Arsenal. The firm undertook much work for the Navy. He was appointed Director General of Ordnance Factories in August 1889.
32 *The Times*, 19/7/1894.
33 *PRI* **13** (1890–92) 7–23; *CN* **61** (1890) 142–25, 147–50; *Rep. BAAS* **61** (1891) 1–33.
34 *RSC* Abel papers, HC 5002/9.
35 *RSC* Abel papers, HC 5002/3.
36 *Hansard*, House of Commons, 25/3/1892, 12/4/1892 and 26/4/1892.
37 *Hansard*, House of Commons, 17/8/1893, 21/8/1893, 24/8/1893; Debate on the Army Estimates, 11 and 12/9/1893.
38 *Final Report*, note 23, p. 130–9.
39 Anderson to Abel, 14/9/1893. *RSC* Abel papers, HC 5002/5.
40 *The Times*, 30/1/1894 – 15/2/1894; The judgment is in *Times Law Reports* **10** (1893–4) 277–80, and is reprinted as Appendix 1, pp. 254–62, in Schück and Sohlman, note 7. The appeal is in *Times Law Reports* **10** (1893–4) 599–600.
41 John Fletcher Moulton (1844–1921) *ODNB*; H.F. Moulton, *The Life of Lord Moulton* (Nisbet, London, 1922). MRI, 1877; FRS, 1880, proposed 4/3/1880, elected 3/6/180; *RSA* EC/1880/13. This quick election at the age of 35, after one paper in *PTRS* on work with William Spottiswoode on electrical discharges through gases, is hard to explain, but may not be unconnected with the fact that Spottiswoode was now President of the Society. Moulton was proposed by Warren de la Rue, supported by Crookes, Rayleigh, Dewar and four other Fellows. He would have known some of these through serving first as a Visitor and then as a Manager of the Royal Institution in 1879–82, and, with de la Rue, he later represented the RI at the International Congress of Electricians in Paris in September 1881; *RIA* MM 5/5/1879, 9/4/1880, 19/7/1881, 3/4/1882. During the First World War he became Director-General of the Explosives Department of the Ministry of Munitions, and a peer.
42 *Otto* v. *Steel*, reported in *The Times*, 15 and 21/12/1885.
43 *Final Report*, note 23, p. 14.
44 E.E. Fournier d'Albe, *The Life of Sir William Crookes* (Fisher Unwin, London, 1923) p. 337.
45 *Times Law Reports* **10** (1893–4) 277–80, and on appeal, 599–600.
46 *Times Law Reports* **4** (1888–9) 686–90, **5** (1889–90) 294–300; *Reports of Patent, Design and Trade Mark Cases* **6** (1890) 243–87, see p. 250.
47 *Times Law Reports* **11** (1894–95) 266–7.
48 Dickson, see Chapter 3, note 18, *RIA* DB 11/1/12a. Dewar himself confirmed this statement in a letter to Kamerlingh Onnes, 19/1/1905.
49 *The Times*, 16/2/1894.
50 H. de Mosenthal, 'The life-work of Alfred Nobel', *J. Soc. Chem. Ind.* **18** (1899) 443–51. Boverton Redwood chaired the meeting.

51 John William Gordon (1853–1936), barrister, admitted to the Middle Temple, 1881, KC 1914; MRI 1899; Hon. Chemical Advisor to the High Explosive Department of the Ministry of Munitions in the First World War; a specialist on patent law; *Who was Who, 1929–1940*.

52 *RIA* MM 10/4/1899.

53 Brock, see Chapter 2, note 40, p. 415.

54 Royal Commission on the Metropolitan Water Supply, House of Commons Parliamentary Papers, C 7172, 1893.

55 Royal Commission, Minutes of Evidence, C7172-1, 1893. Edward Frankland, 21/6/1892, pp. 149–66, and 15/2/1893, pp. 419–24; his son Percy Frankland, 16/11/1892, pp. 419–24; William Odling, 15/11/1892, pp. 378–92; William Crookes, 15/11/1892, pp. 392–6.

56 See Chapter 5, note 49.

57 See Chapter 6, note 52.

58 *RSA* EC/1891/11.

59 Hugo Müller (1833–1915) *ODNB*; German-born industrial chemist who worked with de la Rue and his paper company, a crystallographer and a keen naturalist, H.E. A[rmstrong], *PRS A* **95** (1918–19) xii– xxv. George Matthey (1825–1913) FRS, *ODNB*, a metallurgist. Raphael Meldola (1849–1915) *ODNB*, Professor of Chemistry at Finsbury Technical College, a dyestuffs chemist and amateur naturalist of distinction, a friend of Charles Darwin, W.A. T[ilden] and E.B. P[oulton], *PRS A* **93** (1916–17) xxix–xxxvii. Vernon Harcourt (1834–1919); *DSB*; *ODNB*; Reader in Chemistry at Christ Church, Oxford, and a pioneer in the study of the kinetics of chemical reactions, H.B. D[ixon], *PRS A* **97** (1920) vii–xi.

60 A. Geikie, *Annals of the Royal Society Club* (Macmillan, London, 1917) p. 463.

61 *The Times*, 4/12/1901; Cohen, note 4, pp. 175–9.

62 A.C. Sturney, *The Story of Mond Nickel* (Curwen Press, Plaistow, 1951).

63 L. Mond, 'Metallic carbonyls', *PRI* **13** (1890–2) 668–80; G. Liveing and J. Dewar, 'Note on the spectra of flames of some metallic compounds', *PRS* **52** (1893) 117–23, *CPS* 60. J. Dewar and H.O. Jones, 'Some physical properties of nickel carbonyl', *PRS* **81** (1903) 427–39; *CP* 209. This work was done in 'the winter of 1901'. 'The chemical reactions of nickel carbonyl. Part I, Reactions with halogens and other inorganic substances; Part II, Reactions with aromatic hydrocarbons in the presence of aluminium chloride. Synthesis of aldehydes and anthracene derivatives', *JCS* **85** (1904) 203–12, 212–22; *CP* 214 and 215. 'The physical and chemical properties of iron carbonyl', *PRS* **76** (1905) 558–77; *CP* 226. 'On a new iron carbonyl, and the action of light and heat on the iron carbonyls', *PRS A* **79** (1907) 68–80; *CP* 228. 'Note on carbon monosulphide', *PRS A* (1910) 408–13; *CP* 237. 'The change of carbon disulphide into a gaseous product condensable and explosive near the temperature of liquid air', *PRS A* **83** (1910) 526–9; *CP* 238. 'The interaction of nickel carbonyl and carbon disulphide', *JCS* **97** (1910) 408–13; *CP* 239. 'On the gaseous condensable compound, explosive at low temperature, produced from carbon disulphide vapour by the action of a silent electric discharge', *PRS A* **85** (1911) 574–88; *CP* 240. Jones was killed in the Alps the next year and Dewar abandoned this line of research.

64 British patent No. 27088, 22/12/1898, cited by D. Jeffreys, *Aspirin: The Remarkable Story of a Wonder Drug* (Bloomsbury, London, 2004).

65 A.G. Green, 'Dr Adolph Liebmann', *J. Soc. Dyers & Colourists* **43** (1927) 117. Arthur George Green (1865–1941) was Professor of Colour Chemistry at Leeds, and appeared with Dewar in the Vidal Dyes case, see below.

66 Percy Faraday Frankland (1858–1946), *ODNB*, son of Edward Frankland, and Professor of Chemistry at Birmingham; W.E. Garner, *ONFRS* **5** (1945–48) 697–715.

67 *Reports of Patent, Design and Trade Mark Cases* **22** (1905) 501–18.

68 *The Times* 3/2/1887 and 20/3/1896; 'Reports on Legal Intelligence' in *The Engineer* **81** (1896) 314–15, 426.

69 *Vidal Dyes Syndicate Ltd* v. *Levinstein Ltd and Reed Holliday and Sons*, in the Chancery Court before Mr Justice Parker, 1–12/5/1911 and 16/6/1911. *Reports of Patent, Design and Trade Mark Cases* **28** (1911) 541–63; **29** (1912) 245.

70 Henry Savage Landor, traveller and artist, grandson of Walter Savage Landor, and a personal friend of Dewar. The undated manuscript from which this quotation is taken was written after Dewar's death in 1923 and before Landor's in 1924: *RIA* DB 11/1/18.

Chapter 8 Cryogenics

1 The two related Greek words, κρύος and κρύσταλλος, for *frost* and *ice* respectively, carry implications of both cold and hardness. They have given rise to two now unrelated modern scientific terms, *cryogenic* and *crystal*. The word *cryogen*, a creator of cold, was used for a freezing mixture by Frederick Guthrie in 1875 in the opening volume of his new journal, *Proceedings of the Physical Society*; 'On salt solutions and attached water', *Proc. Phys. Soc.* **1** (1875) 53–101, see 76. The Oxford English Dictionary gives the *Encyclopaedia Britannica* of 1902 for the first use of *cryogenic* to mean related to the science of low temperatures. Although he is not named, the reference is to Dewar's article on 'Liquid gases' in the 10th edition, v. 30, pp. 280–92, see p. 287, where he speaks of a 'cryogenic laboratory'. This was, however, not Dewar's first use of the phrase since it appears in his letter to a Dutch committee enquiring into the safety of the Leiden laboratory in March 1896 (see below). Dewar was well-known to Guthrie, who had been his proposer at his election to the Royal Society in 1877, and he could have obtained the word from Guthrie's use of *cryogen*, but the Dutch origin seems more likely. The phrase 'cryogenic laboratory' had been used by Heike Kamerlingh Onnes in the title of a paper to the Academy of Sciences in Amsterdam in December 1894; 'Over het kryogeen laboratorium te Leiden … .' There was an English translation of this paper a few months later, H. Kamerlingh Onnes, 'On the cryogenic laboratory at Leiden and on the production of very low temperatures', *Comm. Phys. Lab. Leiden*, No. 14 (1895), and the same phrase for the Leiden laboratory was used by a writer using the pseudonym 'Cryos' in May 1896, see note 105 below. The 1895 paper could have been the source of Dewar's use of the word. It had also been used to describe a laboratory in the article on 'Condensation of gases' by J.D. van der Waals in an earlier volume of the same 1902 edition of *Encyclopaedia Britannica*. The etymology and subsequent history of the noun 'cryogenics' are discussed by R.G. Scurlock in his 'Introduction' to the book he edited on *History and Origins of*

Cryogenics (OUP, 1992). He makes no mention, however, of the articles by van der Waals and Dewar. He ascribes the noun 'cryogenics' to the 1940s.

2 J. Dewar, 'Latent heat of liquefied gases', *BAAS Trans. Sect.* **44** (1874) 22, 57; *Nature* **10** (1874) 389, *CP* 42.

3 T. Andrews, 'On the continuity of the gaseous and liquid states of matter', *PTRS* **159** (1869) 575–90.

4 See the Appendix for a brief account of the three phases of matter and the critical state.

5 The early work on refrigeration was well-described in lectures published by James Ewing in 1897, with a second edition in 1908. He took advice from Dewar, who was his colleague in Cambridge, and from the German engineer, Carl Linde, and so his description of the work of these two men does not treat fairly the parallel work of their rival William Hampson, see below. J.A. Ewing, *The Mechanical Production of Cold* (CUP, 1897, 1908).

6 A fuller account of the thermodynamics of gas liquefaction is given in the Appendix.

7 Louis Paul Cailletet (1832–1913); *DSB*. L. Cailletet, 'De la condensation de l'oxygène et de l'oxide de carbone', *CR* **85** (1877) 1213–14. Raoul-Pierre Pictet (1846–1929); *DSB*; R. Pictet, 'Expériences de M. Raoul Pictet sur la liquéfaction de l'oxygène', *CR* **85** (1877) 1214–16, and discussion, 1216–23; L. Cailletet, 'Sur la condensation des gaz réputés incoercibles', *CR* **85** (1877) 1270–1, and discussion, 1271–2.

8 R.-P. Pictet, 'Mémoire sur la liquéfaction de l'oxygène, la liquéfaction, et la solidification de l'hydrogène et sur les théories des changements d'état', *Ann. Chim.* **13** (1878) 145–228.

9 *PRS* **28** (1878–9) 68.

10 *Nature* **17** (1877–8) 169–70, 177–8, 265–9; see also *CN* **37** (1878) 40–1, 83.

11 J. Dewar, 'On the liquefaction of gases', *PRI* **8** (1875–8) 657–63, *CP* 57. W.H. Brock, 'Exploring the Hyperarctic: James Dewar at the Royal Institution', chap. 8 in *James*.

12 *Nature* **18** (1878) 46–7.

13 Zygmunt Florenty von Wróblewski (1845–88); *DSB*, and Notice by Dewar, *CN* **73** (1888) 42; Karol Stanisław Olszewski (1846–1915); *DSB*. Z.F. Wróblewski and K.S. Olszewski, 'Ueber die Verflüssigung des Sauerstoffs, Stickstoffs und Kohlenstoffs', *Ann. Physik* **20** (1883) 243–57, and earlier but informally to the Vienna Academy. (Their published initials change according to the language they were using but Z. and K. are retained here for all their papers.) K. Olszewski, 'On the liquefaction of gases', *Phil. Mag.* **39** (1895) 188–213 is a summary of his own view of his achievements in which he criticised Dewar for ignoring some of his work. Dewar quickly contradicted him on matters of priority in a paper with the same title and published the next month, ibid. 298–305, *CP* 165; Olszewski had overlooked a paper by Dewar with the apparently irrelevant title of 'Recent researches on meteorites', see below, note 31.

14 Z. Wróblewski, 'Dépêche relative à la liquefaction de l'hydrogène', *CR* **98** (1884) 149; K. Olszewski, 'Essais de la liquefaction de l'hydrogène', *CR* **98** (1884) 365–6; 'Température et pression critique de l'azote. Températures d'ébullition de l'azote et de l'éthylène sous faibles presssions', *CR* **99** (1884) 133–6. Wróblewski and Olszewski separated soon after their paper on liquid oxygen in 1883. The cause of the dissolution of their partnership is not known but in his first paper above he

refers formally to Wróblewski as 'mon ancien collaborateur'. They then worked independently until Wróblewski died in an accident in his laboratory in 1888, at the age of 42, when he fell asleep and knocked over a kerosene lamp.

15 *The Times*, 30/6/1891. Dewar to Alfred Nobel, 14/6/86; 'The hole in my neck has been the cause of some rather grave inconveniences and has not yet filled up … a mixture of liquid Oxygen and liquid Ethylene gave a splendid explosion which brought about the disappearance of the apparatus.'

16 Dewar to Kamerlingh Onnes, 20/7/95. I am indebted to Kostas Gavroglu for transcriptions of this and other letters between Dewar and Kamerlingh Onnes, and for his decipherments of the handwriting. The letters are now in the Royal Institution or in the Museum Boerhaave in Leiden.

17 J. Dewar, 'On liquid marsh gas', *Rep. BAAS* **53** (1883) 465, abstracted in *Nature* **28** (1883) 551; *CP* 116.

18 T. O'C. Sloane, *Liquid Air and the Liquefaction of Gases* (Samson Low and Marston, London, 1899) pp. 232–3; for his use of the words 'Dewar bulb', see pp. 244–6. Dewar flasks were not commercially available in Britain at that time. Ramsay was getting his from Berlin in 1899, see Travers, Chapter 5, note 3, p. 189.

19 Z. Wróblewski, 'Sur les propriétés du gaz des marais liquide et sur son emploi comme réfrigérant', *CR* **99** (1884) 136–7.

20 L. Cailletet, 'New process for liquefying oxygen', *Nature* **32** (1885) 584–5.

21 J. Dewar, 'On the critical points of mixed vapours', *PRS* **30** (1880) 538–46, *CP* 83; 'On the lowering of the freezing-point of water by pressure', *PRS* **30** (1880) 533–8, *CP* 82.

22 J. Dewar, 'On the liquefaction of oxygen and the critical volumes of fluids', *PRI* **11** (1884–86) 148–53; *PM* **18** (1884) 210–6; *CP* 122–3; *RIA* MM 3/11/84. J. Dewar, 'On the liquefaction of oxygen and the density of liquid hydrogen', *Rep. BAAS* **54** (1884) 678, abstracted in *Nature* **30** (1884) 351, *CP* 121.

23 J. Dewar, 'Liquid air and the zero of absolute temperature', *PRI* **11** (1885) 318 (title only); *RIA* MM 6/7/85; for Hirst, see Chapter 6.

24 J. Dewar, Friday Evening Discourse of 27 March 1896 on 'New researches on liquid air', *PRI* **15** (1896–8) 133–46, *CP* 170.

25 J. Dewar, 'The American oil and gas fields', *J. Soc. Arts* **33** (1885) 771–80, *CP* 124.

26 Robert Nicol Lennox (1860–1926) [Anon.], *Jour. Proc. Roy. Inst. Chem.* **50** (1926) 296; [Proposal for his election to the Chemical Society], *Proc. Chem. Soc.* **7** (1891) 144. See also, [4th] Lord Rayleigh, 'Some reminiscences of scientific workers of the past generation and their surroundings', *Proc. Phys. Soc.* **48** (1936) 217–46; for Lennox, see 232–5. For Heath, see Chapter 5, note 10.

27 *RIA* MM 3/7/1893 with further payments of £20 on 5/2/1894 and two guineas on 3/12/1894.

28 Sidney Reynolds (1867–1949) was first a Lecturer at University College, London, and then, from 1900 to 1933, Professor of Geology at Bristol.

29 For Marian Reynolds's photograph of the lecturer's bench, 26/6/1891, see *PRI* **13** (1890–2) 481–8, and for her picture of Lennox, see Rayleigh, note 26 above. G.M. Prescott, 'Forging identity: the Royal Institution's visual collections', chap. 4 of *James*.

30 A.H. Eskind, ed., *International Photography, George Eastman House Index to Photographers, Collections and Exhibitions*, v. 3, *Exhibitions* (Prentice-Hall, London, 1998), see *Catalogue of the Special Exhibition of Photography in its*

Application to the Arts, Science and Industry throughout the Empire. Rayleigh had given a Friday Evening Discourse on 23 January 1891 on 'Some applications of photography', *PRI* **13** (1890–2) 261–72.

31 G. Ansdell and J. Dewar, 'On the gaseous constituents of meteorites', *PRS* **40** (1886) 549–59. J. Dewar, 'Researches on meteorites', *PRI* **11** (1886) 541–52, *CP* 128 and 129; Dewar's Friday Evening Discourse of 11 June has the same text as the Royal Society paper but with an additional seven paragraphs on making solid oxygen by evaporating liquid oxygen under reduced pressure, as Stokes reported to the Royal Society on 27 May; *PRS* **40** (1886) 470, *CP* 127; see also *The Times*, 28/5/1886. An abstract of this paper appeared on 19 July in *Industries* **1** (1886) 65, under the title of 'Recent researches on meteorites'.

32 Travers, see Chapter 5, note 3, pp. 188–9.

33 [John] Ambrose Fleming (1849–1945); *DSB*; *ODNB*; W.H. Eccles, *ONFRS* **5** (1945–8) 231–42.

34 Johannes Diderik van der Waals (1837–1923); *DSB*; J.D. van der Waals, *Over de continuiteit van den gas- en vloiestoftoestand* (Thesis, Leiden, 1873), German translation, *Die Continuität des Gasförmigen und Flüssigen Zustandes* (Barth, Leipzig, 1881), English translation, *On the continuity of the gaseous and liquid states*, v. 14 of *Studies in Statistical Mechanics* (North-Holland, Amsterdam, 1988, based on a translation of 1890). A.Ya. Kipnis, B.E. Yavelov and J.S. Rowlinson, *Van der Waals and Molecular Science* (OUP, 1996).

35 J.D.H. Dickson, 'On the criterion by which the critical point of a gas may be determined', *PM* **10** (1878) 40–3.

36 Heike Kamerlingh Onnes (1853–1926); *DSB*; F.G. D[onnan], *PRS A* **113** (1927) i–vi; D. van Delft, *Freezing Physics: Heike Kamerlingh Onnes and the Quest for Cold* (Konink. Nederlandse Akad. Weten., Amsterdam, 2007); [H. Kamerlingh Onnes], *Through Measurement to Knowledge: The Selected Papers of Heike Kamerlingh Onnes, 1853–1926*, edited with an introductory essay by K. Gavroglu and Y. Goudaroulis (Kluwer, Dordrecht, 1991). See also their *Methodological Aspects of the Development of Low Temperature Physics: Concepts out of Context(s), 1881–1956* (Kluwer, Dordrecht, 1989), chap. 2, 'Early research at Leiden and its methodological implication', and K. Gavroglu, 'The reaction of British physicists and chemists to van der Waals' early work and the law of corresponding states', *Hist. Stud. Phys. Biol. Sci.* **20** (1990) 199–237. For the implications for intermolecular forces that Kamerlingh Onnes drew from the law, see J.S. Rowlinson, *Cohesion: A Scientific History of Intermolecular Forces* (CUP, 2002) pp. 186–9.

37 Van der Waals's paper was read to the Academy of Sciences at meetings in October and November 1880 and published in Dutch in the Proceedings ['Investigations of the corresponding properties of equations of state'] *Verhandelingen Konink. Akad. Weten. Amsterdam* **20** (1880) 1–32. Abstracts by Friedrich Roth appeared in German in the *Beiblätter* of the *Annalen der Physik* early in 1881, **5** 27–8, 250–9, and in his German translation of van der Waals's thesis in the same year. Kamerlingh Onnes's paper was in Dutch in the Academy Proceedings, **21** (1881) 8–43, and in French, 'Théorie générale de l'état fluide', *Arch. Néer.* **30** (1881) 101–36, reprinted as Paper 4 by Gavroglu and Goudaroulis, note 36.

38 *RSA* Council Minutes, 24/4/1895.

39 Dewar to Kamerlingh Onnes, 20/7/1895, Kamerlingh Onnes to Dewar, 6/3/1896.

40 He had, however, used liquid nitrous oxide four years earlier; see J. Dewar, 'Magnetic property of liquid oxygen', *PRS* **50** (1891) 247, 261; *PRI* **13** (1889–92) 695–9, *CP* 147–9.

41 Kamerlingh Onnes to Dewar 6/3/1896 and Dewar's reply of 8/3/1896. [On the danger of the use of compressed gases in the Physical Laboratory at Leiden], *Versl. Konink. Akad. Weten. Amsterdam* **5** (1896) 3–17, cited by Gavroglu and Goudaroulis, note 36, p. xli, and by Kipnis et al., note 34, pp. 97–9. *RIA* MM 2/2/1891. The Duchess of Albany was the wife of the youngest son of Queen Victoria; for her visit, see *RIA* MM 10/6/1895.

42 Kamerlingh Onnes to Dewar, 18/7/1895, quoted at greater length by Van Delft, note 36, p. 386; see also R.G. Scurlock, '100 years after Dewar's dewar: cryogenics today', *PRI* **65** (1994) 145–67.

43 George Hugh Gabb (1868–1948); F.S. Taylor, 'Mr. G.H. Gabb', *Nature* **162** (1948) 806, Copy of letter from Gabb to William Bragg, 27 December 1939, Box 6, Gabb papers in the Museum of the History of Science, Oxford.

44 Hermann Johann Phillipp Sprengel (1834–1906) FRS, German-born physicist who worked in London. For Dewar's modification of this pump, particularly for its use in lecture demonstrations, see 'Note on Sprengel's Mercurial Air-Pump', *PRSE* **7** (1872) 662–3, *CP* 19.

45 See Chapter 2, note 38.

46 A.F. Weinhold, *Physikalische Demonstrationen* (Quant & Händel, Leipzig, 1881) pp. 477–80; J. Violle, 'Sur un calorimètre par refroidissement', *CR* **94** (1882) 28–9, 1510–12.

47 Ernest Gaston Joseph Solvay (1838–1922); *DSB*, remembered now for the ammonia-soda process for converting sodium chloride and calcium carbonate into the more valuable sodium carbonate, and for the institution of Solvay Conferences in physics and chemistry.

48 Dewar denied the validity of any claim on behalf of Cailletet in a footnote in his paper, 'The liquefaction of air and research at low temperatures', *PCS* **11** (1895) 221–31, *CP* 169 and in *Nature* **54** (1896) 524. This paper was extended in a Friday Evening Discourse, 'New researches on liquid air', *PRI* **15** (1896) 133–46, *CP* 170. A claim on behalf of Cailletet was repeated in a review of 'Recent Science' by Prince Peter Kropotkin in *The Nineteenth Century* (1898) 259–80 in which he allowed Dewar only the merit of improving the flask. W. Hampson, 'Liquefaction of air by self-intensive refrigeration', *Engineering* **63** (1897) 383.

49 J. Dewar, 'Liquid atmospheric air', *PRI* **40** (1893) 1–12, *CP* 154.

50 W.H. Bragg, 'History of the vacuum flask', two afternoon lectures at the RI, 6 and 13 December 1939, *Nature* **145** (1940) 408–10; [4th Lord] Rayleigh, 'History of the vacuum flask', *Nature* **145** (1940) 625; G.H. Gabb, *Nature* **145** (1940) 862–3; Scurlock, note 42.

51 Copy of a letter from W.J. Green to 4th Lord Rayleigh, 2 May 1940, *RIA* DE 16/2/38.

52 G.D. Liveing and J. Dewar, 'Investigations of the spectrum of Magnesium, No. II', *PRS* **44** (1888) 241–52, *CPS* 55.

53 For Charles Henry Gimingham (1853–1890), the inventor of a multi-tube Sprengel vacuum pump, see *The Electrician* **25** (1890) 625, and for Edward Alfred Gimingham (1866–1953), see *J. Inst. Elec. Eng.* (1953) 237.

54 Copy of a letter from Dewar to Edward Gimingham, 18 December 1892, *RIA* Bragg papers 9 A/2.

55 Reprint from *The Engineer*, 27/1/1893, in *RIA* Bragg papers 9 A/3.

56 The version in *CP* 154 is identical with that in the *PRI*, note 49.

57 Ambrose Fleming to Henry Armstrong, 14/5/1923 in *RIA* D 11/1/4.

58 This version, with its amendments, is taken from Gabb's papers, note 43.

59 The amounts of the payments each year to Edison and Swan, as recorded in the Managers' Minutes at the RI are, as follows: 1886 (£7); 1887 (£21); 1888 (£0); 1889 (£0); 1890 (£0); 1891 (£17); 1892 (£0); 1893 (£32); 1894 (£19); 1895 (£15); 1896 (£27); 1897 (£23); 1898 (£76); 1899 (£26); 1900 (£19); 1901 (£0); 1902 (£0); 1903 (£17).

60 Letters from Dickson to Dewar, from 7/1/1893, in *RIA* D e/8–14. The whole set of Dickson's letters comprise D I e/7 to 59.

61 Experiments made by William Green at Dewar's suggestion, see *RIA* D I 1 h/8.

62 *The Times*, 22/6/1918.

63 J. Dewar, 'The coming of age of the vacuum flask', *PRI* **21** (1914–16) 240–62, *CP* 245; A. Room, *Dictionary of Trade Name Origins* (Routledge and Kegan Paul, London, 1982) p. 169.

64 *RIA* D II i/3.

65 T. Andrews 'On a method of obtaining a perfect vacuum in the receiver of an air-pump', *PM* **3** (1852) 104–8; Andrews cannot be blamed, however, for Tait and Dewar's solecism of 'very perfect vacua'. For their work, see Chapter 2, note 39.

66 This subject was discussed in a paper at the BAAS meeting at Cambridge in 1904 where it is represented by title only, 'On new low-temperature phenomena and their scientific applications', but of which a substantial abstract was published in *The Times*, 25/8/1904; *CP* 221.

67 J. Dewar, 'New low temperature phenomena', *PRI* **18** (1905) 177–97; *CP* 223.

68 J. Dewar, 'Experiments in the liquefaction of hydrogen', *CN* **70** (1894) 115–16, reprinted from *The Times*, 1/9/1894; *CP* 258.

69 C.W. Siemens's report of 1857 was printed in 1882 in a discussion of a paper by Joseph Coleman to the Institution of Civil Engineers; J.J. Coleman, 'Air refrigeration machinery and its applications', *Min. Proc. Inst. Civil Eng.* **68** (1882) 146–69, discussion, 170–215. Joseph Coleman (1828–88) was an English industrial chemist who worked in Scotland at James Young's oil and gas fields where he became involved in the problems of refrigeration. He is remembered now for the Bell-Coleman cycle for the refrigeration of foodstuffs.

70 Alexander C. Kirk (1830–92), a Scots engineer who worked in the refrigeration, oil shale and shipbuilding industries. His refrigeration work is described in a paper and its subsequent discussion at the Institution of Civil Engineers, 'On the mechanical production of cold', *Min. Proc. Inst. Civil Eng.* **37** (1873–4) 244–315; see also J.A. Ewing, *The Mechanical Production of Cold* (CUP, 1908) pp. 34–8.

71 J.J. Coleman, 'The mechanical production of cold and the effect of cold on microphytes', *PRI* **11** (1884–6) 148–53. Both here and in his paper to the Civil Engineers, note 69, he mentioned briefly the Joule and Joule-Thomson effects but most of the RI discourse was on the use of cold in preventing the putrefaction of meat.

72 Solvay recalled his work of 10 years' earlier in a paper of December 1895, 'Sur la production mécanique des températures extrêmes', *CR* **121** (1895) 1141–3. He had

then just learnt of Linde's success in using an interchanger, and extracting the heat by expansion through a valve that had no moving parts; see below.

73 H. Chang, *Inventing Temperature* (OUP, 2004) pp. 182–97; H. Chang and S.W. Yi, 'The absolute and its measurement: William Thomson on temperature', *Ann. Sci.* **62** (2005) 231–45; J.S. Rowlinson, 'James Joule, William Thomson and the concept of a perfect gas,' *NRRS* **64** (2009) 43–57.

74 The three methods of expansion can be described in thermodynamic terms as isentropic, isoenergetic and isenthalpic, that is, they measure the derivatives $(\partial T/\partial p)_S$, $(\partial T/\partial p)_U$ and $(\partial T/\partial p)_H$ respectively. The calculations of these in the text are only approximations based on the perfect gas laws for the first calculation and on the measured values of the second virial coefficient for the second and third. In the Appendix there is a fuller analysis of the first expansion, as applied in Cailletet's original experiment, and of the third, or Joule-Thomson, expansion.

75 [Lord] Rayleigh, 'Liquid air in one operation', *Nature* **58** (1898) 199. Rayleigh's proposal may have been behind a patent from an engineer, Edward Thrupp, 'Improvements in apparatus for refrigeration and liquidation of air and other gases', British patent 26767 (1898). Thrupp seems never to have exploited it. See also, G. Claude, 'Sur la liquefaction de l'air détenté avec travail extérieur récupérable', *CR* **130** (1900) 500–2, **134** (1902) 1568–70. Such a method of cooling was tried in the 1930s by the Linde Company but its first successful practical use seems to be by Piotr Kapitza in Moscow for the liquefaction of helium in 1939; see J.G. Daunt, 'The production of low temperatures down to hydrogen temperatures', in S. Flügge, ed., Handbuch *der Physik* (Springer, Berlin, 1956) chap. 1.

76 Mansel Davies disinterred everything that seems to be known about Hampson, whose name is in *DSB* but not in *Who was Who* nor in *ODNB*; see M. Davies, 'William Hampson (1854–1926)' *BJHS* **22** (1989) 63–73. After Hampson's work on gas liquefaction had passed into the hands of Ramsay and Travers at University College, London, he worked on electrical and other medical physics devices in the London hospitals.

77 W. Hampson, 'Liquefaction of hydrogen', *Nature* **58** (1898) 77.

78 *RIA* MM, 3/12/1894.

79 W. Hampson, 'Liquefaction of air', *Engineering* **65** (1898) 377; 'Arenal', *Engineering* **65** (1898) 396; Hampson, *Engineering* **65** (1898) 474; 'Arenal', *Engineering* **65** (1898) 508. In this letter 'Arenal' introduced into the argument a patent of 1893 by an American engineer, Charles E. Tripler, which the editor of the journal satd, in a footnote, is clearly not an anticipation of Hampson and Linde's patents, and which Hampson dismissed as irrelevant, *Engineering* **65** (1898) 541–2. (Tripler's claim was put forward again the next year by Sloane, note 18, whose book received a scathing review, probably by Hampson, in *Nature* **60** (1899) 268–9.) Hampson then points out that 'Arenal' is Lennox, whose initials are R.N.L.; Hampson, *Engineering* **65** (1898) 573; Lennox, *Engineering* **65** (1898) 607, said that Hampson should be grateful to him for benefit of his advice in 1894!

80 The provisional patent application is headed: 'William Hampson, MA, LSA, of Bryn, Nannerch, Mold, Chester, and 334 City Road, EC, no occupation, "Improvements relating to the progressive refrigeration of gases": Provisional application, 23 May 1895, No. 101325; Complete Application, 19 February 1896.' The punctuation of his address is wrong; 'Bryn Nannerch' was presumably his family home. It is a fine house which still stands in the main street of Mold in the Welsh

county of Flintshire, but his working life was spent in London. LSA = Licentiate of the Society of Apothecaries, the qualification that allowed him to work in hospitals on problems of medical physics, including the newly discovered x-rays.

81 The Brin Company was founded by two French chemists, Quentin and Arthur Brin, who operated their process in Westminster under their British patent of 1886. The name was changed to British Oxygen Company in 1906. See W.B. Jensen, 'The origin of the Brin process', *J. Chem. Educ.* **86** (2009) 1266–7.

82 *RIA* DB I/8 p. (49) or 62. This notebook in Dewar's handwriting has two sets of page numbers which differ by 13. One set, apparently the original, is in parentheses; only this is shown here. The successive entries mentioned in the text are, as follows: 6 August 1894, p. (49); undated, ca. June 1895, p. (89); 5 August 1895, p. (95); undated, pp. (130–1).

83 *RIA* MM 2/12/1895; *The Times*, 3/12/1895.

84 Hugh Munro Ross (1870–1954), obituary in *The Times*, 9/9/1954.

85 W. Hampson, 'Liquid air', *Engineering* **60** (1895) 743; *The Engineer* **80** (1895) 617.

86 J. Dewar, 'Liquid hydrogen', *Nature* **58** (1898) 125.

87 J. Dewar, 'The liquefaction of air and research at low temperatures', *PCS* **11** (1896) 221–31, and discussion, 231–4; *CN* **73** (1896) 40–6. This lecture was repeated in an extended and slightly emended form as a Friday Evening Discourse at the RI, note 24; *CP* 169–70.

88 J. Dewar, 'Liquid hydrogen', *Nature* **58** (1898) 199.

89 E. Blass, 'A problem in thermodynamics', *Nature* **52** (1895) 415; E.T. Dixon, *Nature* **52** (1895), 547.

90 Carl Linde (1842–1934); *DSB*, was a German engineer who had two periods of teaching at Munich Polytechnic. Between these periods he established a successful company for making refrigeration plant, Gesellschaft für Linde's Eismaschinen. This company had a branch in London, Linde British Refrigeration Co., which advertised regularly in the British engineering press.

91 M. Schröter, 'The commercial manufacture of liquid air and oxygen', *The Engineer* **80** (1895) 325.

92 Dewar to Kelvin, 29/10/1895, *CUL* Add. Ms. 7342, D 87.

93 C. Linde, 'Professor at Munich, Process and Apparatus for liquefying gases or gaseous mixtures, and for producing cold, more particularly applicable for separating oxygen from atmospheric air', Patent No. 12526, Provisional Application, 28 June 1895, Complete Application, 25 March 1896. The German patent of 1895 [not seen] is No. 88824.

94 *RIA* D I d/19.

95 J. Dewar, 'The scientific work of Joule', *PRI* **13** (1890–2) 1–6, *CP* 144. The Discourse was given on 24 January 1890.

96 Dickson to Dewar, 20/8/1879, D I e/7.

97 Dickson to Dewar, 11/12/1895, *RIA* D Vb/2.

98 [Editorial], *The Engineer* **80** (1895) 557, 6/12/1895.

99 [Editorial], *Engineering* **61** (1896) 263; R.N. Lennox, *Engineering* **65** (1898) 542.

100 H.D. Richmond, 'Bertram Blount (1866–1921)', *Analyst* **46** (1921) 269–70.

101 'The production of liquid air', *The Engineer* **80** (1895) 644; **81** (1896) 21.

102 'X', 'Regenerative cold', *The Engineer* **81** (1896) 91–2.

103 'A new process for the liquefaction of air and other gases', *Nature* **53** (1896) 515–16.

104 'Zero', 'The Hampson liquid oxygen process', *The Engineer* **81** (1896) 326.

105 C.E. Basevi, 'The Joule[-Thomson] effect', *The Engineer* **81** (1896) 550–1. The letters from 'Cryos', 'Zero', 'J' and 'MA' do not need here more than a bare list of their volume and page numbers: **81** (1896) 67, 120, 326, 372, 396, 422, 467, 493, 516, 566–7.

106 C. Linde, 'Process and Apparatus for attaining lowest temperatures, for liquefying gases, and for the mechanical separation of gas mixtures', *The Engineer* **82** (1896) 485–6, 509.

107 C. Linde, 'Zur Geschichte der Maschinen für die Herstellung flüssiger Luft', *Ber. Deut. Chem. Ges.* **32** (1899) 925–7. Linde's words were: 'und mit keine Worte erwähnt dieselbe eine Anwendung der durch Thomson und Joule gefunder Ausströmungs-Abkühling als Kälte Quelle'.

108 W. Hampson, 'Self-intensive refrigeration of gases: liquid air and oxygen', *Jour. Soc. Chem. Ind.* **17** (1898) 411–17, followed by a contribution from Dewar and further discussion, 417–21.

109 For a report of the announcement of Dewar's liquefaction of hydrogen, and for Hampson and Dewar's subsequent letters, all entitled 'Liquid hydrogen', see *Nature* **58** (1898) 55–7, 77, 125, 174, 199, 246, 270, 292 and 319.

110 *The Times*, 1/9/1894.

111 *RI* MM 14/6/1897. Moissan to Dewar, 19/5/1897 and 3/6/1897, D II c 18, 19. In his second letter Moissan reports that he had now liquefied fluorine in Paris on his return from London.

112 H. Moissan and J. Dewar, 'Sur la liquéfaction du fluor', *CR* **124** (1897) 1202–5; 'Nouvelles experiences sur la liquéfaction de fluor', *CR* **125** (1897) 505–11; 'On the properties of liquid fluorine', *PCS* **13** (1897) 175–84, *CP* 186, and, by title only, *Rep. BAAS* **67** (1897) 611; See also H. Moissan and J. Dewar, 'Sur la solidification du fluor et sur la combinaison à -252,5 du fluor solide et de l'hydrogène liquide', *CR* **136** (1903) 641–3, *CP* 210. *RIA* D II c/18 and 19.

113 The rough sketch is in his laboratory notebook *RIA* D V b/1 and has been reproduced by K. Gavroglu, 'On some myths on the liquefaction of hydrogen and helium', *Eur. Jour. Phys.* **15** (1994) 9–15. See also, K. Gavroglu, 'A pioneer who never got it right: James Dewar and the elusive phenomena of cold', in *Going Amiss in Experimental Science*, Boston Studies in the Philosophy of Science, **209** (2009) 137–57. I have generally followed Gavroglu's transcription and decipherment of Dewar's entries in his laboratory notebook, although with a few differences arising from our interpretions of the handwriting.

114 J. Dewar, 'Gases at the beginning and end of the century', *PRI* **16** (1901) 730–7, *CP* 199.

115 J. Dewar, 'Liquid hydrogen', *PRI* **16** (1899) 1–14, *CP* 193. This Discourse contains an edited version of the hasty first announcements and omits the claims to have liquefied helium. These announcements of his success were to *The Times*, 11 May 1898, then to the Royal Society, on 12 May, 'Preliminary Note on the liquefaction of hydrogen and helium', *PRS* **63** (1898) 231, 256–8, followed, on the 15 December, by 'On the boiling-point of liquid hydrogen under reduced presssure', *PRS* **64** (1899) 227–51. His communications to the Chemical Society were on 19 May, 'Notes on the liquefaction of hydrogen and helium', *JCS Trans.* **73** (1898) 528–35,

and on 2 June, 'The boiling point and density of liquid hydrogen', *PCS* **14** (1898) 146–7; see also note 110. His papers generally contained thanks to Lennox and Heath but that to Lennox at the end of the RI Discourse is unusually generous: 'it is not too much to say that but for his engineering skill, manipulative ability and loyal perseverance, the present successful issue might have been indefinitely delayed'. Lennox's and Heath's salaries were raised the next month, *RIA* MM 6/6/1898.

116 Dewar to Kamerlingh Onnes, 11/6/1898, quoted by Gavroglu, note 113 (1994).

117 J. Dewar, 'Solidification of hydrogen', *Rep. BAAS* **69** (1899) 683; 'Solid hydrogen', *PRI* **16** (1900) 473–80, *CP* 196–7.

118 J. Dewar, 'Liquid air as an analytic agent', *PRI* **15** (1898) 815–29; 'Liquid hydrogen' (Centenary Lecture), *PRI* **16** (1899)1–14. *CP* 189 and 193. Rayleigh's lecture was summarised in 'Centenary of the Royal Institution', *Nature* **60** (1899) 129–30.

119 G.D. Liveing and J. Dewar, 'On the spectrum of the more volatile gases of atmospheric air, which are not condensed at the temperature of liquid hydrogen – Preliminary Notice', *PRS* **67** (1900) 467–74, *CPS* 71. J. Dewar, 'The boiling point of liquid hydrogen as determined by a rhodium-platinum resistance thermometer', *PCS* **15** (1899) 70–1, *CP* 194.

120 J. Dewar, 'Boiling point of liquid hydrogen, determined by hydrogen and helium gas thermometers', *PRS* **68** (1901) 44–54, *CP* 200.

121 *RIA* D V b/1, p. (17) 26. This notebook also has two sets of numbering of the pages; cf. note 82.

122 *RIA* MM 7/5/1923.

123 See Chapter 6, note 52.

124 [Anon.], 'Tribute to Hertz', *Nature* **50** (1894) 148; *RIA* MM 6/1/1899.

125 Redwood to Dewar, 24/11/1903; *RIA* D VI/2; DE 7/8/1.

126 R.S. Hudson, *Recollections of a Technologist* (Pitman, London, 1964) p. 45.

127 Joseph Ernest Petavel (1873–1936) was London-born but of Swiss descent. He was assistant to Ambrose Fleming at University College, London and, on being awarded an '1851 Exhibition', became one of the first two scientists admitted to the Davy Faraday Laboratory when it opened in January 1897; see Chapter 10. He left for Manchester University after a row with Dewar but was readmitted so he could take charge of the liquid hydrogen apparatus at the St Louis Exhibition in the summer and autumn of 1904, *RIA* MM 7/12/1903. He ended his career as Director of the National Physical Laboratory. *ODNB*; R. Robertson, *ONFRS* **2** (1936–1938) 183–213.

128 Undated manuscript of a popular lecture, *RIA* D II b/61.

129 Lord Kelvin, 'Nineteenth century clouds over the dynamical theory of heat and light', *PM* **2** (1901) 1–40, a Friday Evening Discourse on 27/4/1900.

130 Dewar, note 120, *CP* 200; 'The nadir of temperature and allied problems' (Bakerian Lecture) *PRS* **68** (1901) 360–6, *CP* 203; 'On electrical resistance thermometry at the temperature of boiling hydrogen', *PRS* **73** (1904) 244–51, *CP* 216; 'Heat problems', *PRI* **20** (1912) 577–90, *CP* 243; J.D.H. Dickson, 'On the reduction to normal air temperatures of the platinum-temperatures in the low-temperature researches of Professors Dewar and Fleming', *PM* **45** (1898) 525–8, *CP* 264; J.A. Fleming, 'Electrical and magnetic research at low temperatures', *PRI* **15** (1896) 239–68, *CP* 263.

131 J.D.H. Dickson, 'The thermo-electric diagram from -200 °C to 100 °C, deduced from the observations of Professors Dewar and Fleming', *TRSE* **47** (1911) 737–91, *CP* 265.

132 G.D. Liveing and J. Dewar, 'Notes on the absorption-spectra of oxygen and some its compounds', *PRS* **46** (1889) 222–32, *CPS* 57; see also, G.D. Liveing and J. Dewar, 'On the spectrum of liquid oxygen, and of the refractive indices of liquid oxygen, nitrous oxide and ethylene', *PM* **34** (1892) 205–9, and *Rep. BAAS* **62** (1892) 658; *CPS* 61.

133 J. Dewar, 'Application of liquid hydrogen to the production of high vacua, together with their spectroscopic examination', *PRS* **64** (1898) 231–8, *CPS* 70; J. Dewar and G.D. Liveing, 'On the spectrum of the more volatile gases of atmospheric air which are not condensed at the temperature of liquid hydrogen – Preliminary Notice', *PRS* **67** (1900) 467–71, *CPS* 71; J. Dewar, 'The separation of the most volatile gases from air without liquefaction', *PRS* **74** (1904) 127–31, *CPS* 77.

134 Dewar to William Thomson, 10/12/1891 and 17/12/1891, *CP* 147–8; J. Dewar, 'Magnetic properties of liquid oxygen', *PRI* **13** (1892) 695–9, *CP* 149.

135 Ambrose Fleming to Henry Armstrong, 14/5/1923, *RIA* DB 11/1/4.

136 The work of Dewar and Fleming was published in a series of papers, first in the Philosophical Magazine and then in the Proceedings of the Royal Society. It needs no detailed description; the papers are in *PM* **34** (1892) 205–9; **40** (1895) 95–119, 303–11; *PRS* **60** (1896) 285–96, 358–68, 425–32; **61** (1897) 2–18, 299–316, 316–30, 358–67, 368–80, 381–96; **62** (1897) 250–66; **63** (1898) 311–29; *CP* 152, 166, 167, 175, 176, 178–85, 190.

137 He had long had an interest in calorimetry but this series of experiments started, in effect, with J. Dewar, 'Liquid hydrogen calorimetry', *PRI* **17** (1904) 581–96, 'New low temperature phenomena', *PRI* **18** (1905) 177–92, and culminated in J. Dewar, 'Atomic specific heats between the boiling-points of liquid nitrogen and liquid hydrogen; I, The mean atomic specific heats at 50° abs. of the elements, a periodic function of their atomic weights', *PRS A* **89** (1913) 158–69; *CP* 218, 223 and 244, see also *CP* 256–7.

138 The slow appreciation by both physicists and chemists of Einstein's quantal theory of the specific heat of simple solids is discussed by T.S. Kuhn, *Black-body Theory and the Quantum Discontinuity, 1894–1912* (OUP, 1978) pp. 210–20. Dewar's reproduction of Einstein's curve for a specific heat is in his paper on 'Heat problems', *PRI* **20** (1912) 579–90, *CP* 243.

139 Pierre Curie (1859–1906) *DSB*. P. Curie, 'Le Radium', *PRI* **17** (1903) 389–402.

140 P. Curie and A. Laborde, 'Sur la chaleur dégagée spontanément par les sels de radium', *CR* **136** (1903) 673–5; see also *RIA* D I d/36, 6/7/1903. W. Crookes and J. Dewar, 'Note on the effect of extreme cold on the emanations of radium', *PRS* **72** (1903) 69–71; *CP* 212. This was Crookes's only joint paper.

141 [J.] Dewar and [P.] Curie, 'Examen des gaz occlus ou dégagés par la bromure de radium', *CR* **138** (1904) 190–2; J. Dewar, 'Note on the use of the radiometer in observing small gas pressures; Application to the detection of the gaseous products produced by radio-active bodies', *PRS A* **79** (1907) 529–32; 'The rate of production of helium from radium', *PRS A* **81** (1908) 280–6; 'Problems of helium and radium', *PRI* **19** (1909) 724–34; 'Long-term determination of the rate of production of helium from radium', *PRS A* **83** (1910) 404–8; *CP* 213, 230, 233, 234, 236.

Chapter 9 Argon and Helium

1 Rayleigh, 'On the relative densities of hydrogen and oxygen', *Nature* **46** (1892) 101–4; 'Density of nitrogen', *Nature* **46** (1892) 512–13. In a later paper read at the Royal Society on 19 April 1894, Rayleigh set out in detail the work on the densities of the two kinds of nitrogen but without making any suggestion of the origin of the differnce; 'On an anomaly encountered in determinations of the density of nitrogen gas', *PRS* **55** (1894) 340–4.

2 Ramsay to Rayleigh, 20/11/1892, quoted by Strutt, see Chapter 3, note 14, p. 189–90.

3 H. Cavendish, 'Experiments on air', *PTRS* **75** (1785) 372–84, see 382. The paper was read on 2 June 1785, and is sometimes confused with an earlier paper of the same title read the previous year. Both Dewar's and Ramsay's immediate source was the biography; G. Wilson, *The Life of the Hon. Henry Cavendish* (Cavendish Society, London, 1851) p. 259.

4 Strutt, see Chapter 3, note 14, pp. 193–4.

5 Kelvin to Stokes, 9/3/1896, printed by D.B. Wilson, *The Correspondence between Sir George Gabriel Stokes and Sir William Thomson, Baron Kelvin of Largs* (CUP, 1990) Letter no. 553.

6 Strutt, see Chapter 3, note 14, p. 203.

7 J. Dewar, 'The new element', *The Times*, letters of 15 and 16/8/1894, published 16 and 18/8/1894.

8 *RIA*, T.A. Hirst, *Journal*, 25/11/1891; N. Tesla, 'Alternating currents of high potential and high frequency', *PRI* **13** (1893) 637–52.

9 Dewar to Kelvin, 23/11/1894, Add Ms. 7342, D 86, in D.B. Wilson, *Catalogue of the Manuscript Collections of Sir George Gabriel Stokes and Sir William Thomson, Baron Kelvin of Largs in the Cambridge University Library* (CUP, 1976). The experiments were made on 21/11/1894, *RIA* DB I pp. (54–66), or 67–79, since each page carries two numbers. The original ones, in parentheses, are used here.

10 J. Dewar, 'The relative behaviour of chemically prepared nitrogen and of atmospheric nitrogen in the liquid state', *PCS* **10** (1894) 222–5; *CP* 163.

11 Dewar to Rayleigh, 4/12/1894, 6/12/1894 and Rayleigh to Dewar, 5/12/1894, cited by Strutt, Chapter 3, note 14, pp. 212–13. Wyndham Dunstan was a pharmaceutical chemist who was at the Imperial Institute, where the Director, whom he was to succeed, was Frederick Abel, Dewar's collaborator in the discovery of cordite. It may have been through Abel that Dewar knew Dunstan, or through their meeting at the Chemical Society where Dunstan was a Secretary. Strutt's biography was published in 1924 when Dunstan was still alive and his name was suppressed, but it was added in the revised edition published in 1968 by the University of Wisconsin Press, after Strutt's own annotated copy had come to light.

12 'The New Element', *The Times*, 7/12/1894, cited in part by Strutt, Chapter 3, note 14, pp. 213–14.

13 *New York Times*, 6/8/1895.

14 RS Council Minutes, 5/7/1894 and 1/11/1894.

15 Lord Rayleigh and W. Ramsay, 'Argon, a new constituent of the atmosphere', *PRS* **57** (1894–5) 265–86; W. Crookes, 'On the spectrum of argon', *PRS* **57** (1894–5) 287–9; K. Olszewski, 'The liquefaction and solidification of argon', *PRS* **57** (1894–5) 290–2

16 'Argon', *Nature* **51** (1894–5) 337–8; 347–56.
17 Strutt, Chapter 3, note 14, p. 222.
18 'The Chemical Society', *The Times*, 28/3/1895.
19 The number 587.49 (now reckoned to be 587.562) is the wave-length of the D_3 line of the spectrum of helium, in nanometres. The better known doublet of sodium, D_1 and D_2, is at 588.995 and 589.592 nm. The fullest accounts of the discovery of helium and the other inert gases are those given by Morris Travers in *The Experimental Study of Gases* (Macmillan, London, 1901) chap. 11, 15 and 16; *Discovery of the Rare Gases* (Arnold, London, 1928) pp. 56–8, 65 and 99; and in chap. 8 of his biography of Ramsay, see Chapter 5, note 3. W.H. Keesom, Kamerlingh Onnes's successor as Joint Director of the Leiden Laboratory, wrote the classic work on helium during the Second World War; *Helium* (Elsevier, Amsterdam, 1942).
20 For Rayleigh, see *The Times*, 18/1/1896. For Dewar, see Brock, Chapter 2, note 40, p. 348. For Mond, see Cohen, Chapter 7, note 4, pp. 225–7, who says that it was when Dewar and his wife were at the Mond's Christmas party at the Landsdown Hotel in Bath that Mond collected some of the Bath gas and passed it to Dewar for analysis. Ramsay asked Arthur Richardson, his former colleague at Bristol, to collect some gas for him, see Travers, Chapter 5, note 3, p. 129.
21 J. Dewar, 'Liquid air as an analytic agent', *PRI* **15** (1898) 815–29, *CP* 189.
22 P. Ford and R. Rolls, 'Airs and waters: the hot springs and Bath chemists', chap. 1 of P. Wallis, ed., *Invention and Discovery: Bath and the Rise of Science* (Bath Royal Literary and Scientific Society and William Herschel Society, Bath, 2008); *RIA* MM 11/6/1900, 5/11/1900, 4/11/1901, 3/11/1902, 6/4/1903.
23 *RIA* MM, 7/5/1906.
24 At the time of their discovery this group of gases was usually called the 'rare gases' or the 'inert gases'. Neither name is now thought appropriate since helium is the second commonest element in the universe, argon is far from rare, and xenon, in particular, is not totally inert. The International Union of Pure and Applied Chemistry has given its imprimatur to the name 'noble gases' but few yet use it and here I retain 'inert' as the most common name in Ramsay and Dewar's time.
25 Dickson to Dewar, 7/6/1900, *RIA* D1 e/23. By 'exhaustion' Dickson meant evaporation by pumping to reduce the vapour pressure and so the temperature of the boiling point.
26 J. Dewar, Presidential Address, *Rep. BAAS* **72** (1902) 1–47, *CP* 207.
27 For careful discussions of the principles involved, see A.B. Pippard, *The Elements of Classical Thermodynamics* (CUP, 1961) pp. 48–52; H. Buchdahl, *The Concepts of Classical Thermodynamics* (CUP, 1966) chap. 7.
28 *RIA* D V/b, p. (52 et seq.); Dewar, Chapter 8, note 130, Bakerian Lecture; 'Nouvelles recherches sur la liquéfaction de l'hélium', *CR* **139** (1904) 421–2.
29 H. Kamerlingh Onnes, 'Experiments on the condensation of helium by expansion', *Proc. R. Acad. Sci. Amsterdam* **10** (1908) 744–7; *Comm. Phys. Lab. Leiden*, No. 105 (1908).
30 M.W. Travers and A. Jaquerod, 'On the vapour pressure of liquid hydrogen at temperatures below its boiling-point on the constant volume hydrogen and helium scales', *PTRS A* **200** (1903) 155–76. Hampson had designed and Travers had built an apparatus for liquefying hydrogen at University College in 1900.
31 *RIA* MM 5/2/1900, 4/11/1901; Dickson to Dewar, 1/9/1901, *RIA* DE 10/45/41.

32 *RIA* D V/b, p. (107) and 119. K. Gavroglu, 'James Dewar's nemesis: the liquefaction of helium', *PRI* **65** (1994) 169–85.
33 Dewar to Kamerlingh Onnes, 7/5/1903.
34 Dickson to Dewar, 11/12/1902, *RIA* D 1e/31.
35 H. Kamerlingh Onnes and W.H. Keesom, *Comm. Phys. Lab. Leiden* (1906) No. 96a, 96b, and 96c.
36 J. Dewar, 'Studies in high vacua and helium at low temperatures', *PRS* **18** (1907) 747–56, *CP* 229.
37 Dewar to Kamerlingh Onnes, 5/1/1904.
38 Dewar to Kamerlingh Onnes, 12/7/1905. Gavroglu notes that the phrases in italics are in the draft of the letter in Dewar's papers but not in the letter as received at Leiden. Kamerlingh Onnes to Dewar, 29/12/1903.
39 H. Moissan and J. Dewar, 'Sur la solidification du fluor et sur la combinaison à -252.5 du fluor solide et de l'hydrogène liquide', *CR* **136** (1903) 641–3; 'Sur l'affinité à basse température; réactions du fluor à -187', *CR* **136** (1903) 785–8; *CP* 210–11; with Crookes, and with Curie, Chapter 8, notes 140 and 141.
40 W.J. Green, 'Some Memories of the Royal Institution and its Laboratories, 1900–1950', *PRI* **36** (1956) 543–67, reprinted as Appendix 2 in *James*. See also the view of the 4th Lord Rayleigh in Chapter 8, note 26.
41 Dewar to Scott, 15/11/1908, *RIA* DC 2/45.
42 'Solidification of helium', *The Times* 6/3/1908, 11/3/1908.
43 Kamerlingh Onnes to Dewar, 7/4/1908. Dewar in *The Times* on the reported solidification of helium, 16/4/1908.
44 Dewar to Kamerlingh Onnes, 15/4/1908.
45 'The liquefaction of helium', *The Times*, 17/7/1908.
46 [H.] Kamerlingh Onnes, 'The liquefaction of helium' followed by J. Dewar 'Anticipations and experiments on the liquefaction of helium' (both by title only), *Rep. BAAS* **78** (1908) 673. Kamerlingh Onnes to Dewar, 25/9/1908; Dewar to Kamerlingh Onnes, 29/9/1908. Van Delft, see Chapter 8, note 36, pp. 460–5.
47 J.C. McLennan, 'Electrical conductivity of metals at the lowest temperature', *PRI* **27** (1932) 446–70; R. Fox and G. Gooday, eds, *Physics at Oxford, 1839–1939* (OUP, 2005) pp. 245–6.
48 Royal Society Referees' Reports, v. 9, no. 389, 14/8/1886.
49 Ramsay's first claim of the liquefaction of hydrogen by Olszewski is in the discussion of Dewar's paper in which he introduced his new apparatus for the liquefaction of oxygen, *PCS* **11** (1895) 231–4, *CP* 169.
50 Ramsay's second claim for Olszewski's priority, in 1898, was dismissed at length by Dewar his Friday Evening Discourse on the liquefaction of hydrogen, J. Dewar, 'Liquid hydrogen', *PRI* **16** (1899) 1–14, *CP* 193.
51 *The Jubilee of the Chemical Society* (Chemical Society, London 1896). The major debate among British physical chemists at the end of the nineteenth century was that on the nature of solutions of electrolytes. Dewar seems to have stayed aloof from this, but it may have contributed to the mutual opposition of Armstrong and Ramsay, since Pattison Muir, Rayleigh and Ramsay backed the views of Ostwald and Arrhenius against the opposition of Armstrong who never accepted the reality of an ionic constitution of acids, bases and salts; see R.G.A. Dolby, 'Debates over the theory of solution: a study of dissent in the English-speaking world in the late nineteenth and early twentyith centuries', *Hist. Stud. Phys. Sci.* **7** (1976) 297–404.

52 The Chemical Society election was described in detail in successive issues of *CN* **75** (1897) 139, 154–5, 162–3, 166, 167, 178–9, 190, 201, 210–12 259 and 260.
53 The phrase 'Suum Cuique' comes from Cicero and may be translated as 'to each his own', or 'everyone gets his true deserts'. It acquired a bad reputation after the Second World War since its German translation, 'Jedem das Seine' had been inscribed over the entrance to the concentration camp at Buchenwald.
54 'Suum Cuique', 'Gas and gases', *CN* **77** (1898) 284. The paper of Liveing and Dewar to which 'Suum Cuique' refers is 'Preliminary note on the spectrum of the electric discharge in liquid oxygen, air, and nitrogen', *PM* **38** (1894) 235–40, *CPS* 64, although it is difficult to see much evidence there for 'Suum Cuique's' claim.
55 J. Dewar, 'Metargon', *CN* **78** (1898) 70–1. Metargon had been proposed and named by Ramsay and Travers, 'On the companion of argon', *CN* **78** (1898) 1–2.

Chapter 10 The Davy Faraday Research Laboratory

1 A full account of the founding and the operation of this laboratory in Dewar's time is given by K.D. Watson in chapter 9 of *James*, '"Temporary Hotel Accommodation"? The Early History of the Davy-Faraday Research Laboratory, 1894–1923'. Here I use her account, but consider in full only the part played by Dewar in this unusual annexe to the RI. In the Deed of Trust of 8 June 1896 the title of the Laboratory was as given here; the insertion of a hyphen to give 'Davy-Faraday' came later.
2 *RIA* MM 2/4/1894.
3 *RIA* MM 7/5/1894.
4 *RIA* MM 10/4/1893.
5 *RIA* MM 2/7/1894.
6 *RIA* MM 1/4/1895.
7 *RIA* MM 5/11/1894, 6/5/1895.
8 *RIA* MM 4/11/1895, 12/6/1896.
9 John Shields (1869–1920) was a link between the DFRL and Ramsay's team at University College. He had been educated at Edinburgh and Strassburg, worked briefly with Ostwald and Arrhenius, and then took his 1851 Exhibition to University College to work with Ramsay on the surface tension of liquids. From there he went to Mond's private laboratory and so to the DFRL in 1897, perhaps as Mond's nominee. He was at first refused entry because his proposed project was too close to Dewar's previous work on the absorption of hydrogen in palladium. He was admitted after a second choice of topic. He ended his career as Chief Chemical Engineer with the Rio Tinto mining company; see J[ames] W[alker] *JCS* **119** (1921) 569–71.
10 *RIA* MC DFRL, 26/6/1896, 3/7/1896, 27/10/1896.
11 *Illustrated London News*, 2/1/1897, see Watson, note 1, p. 198.
12 Caroe, see Chapter 6, note 1, p. 94.
13 *RIA* MM 1/6/1897; MC DFRL, 30/10/1900.
14 Petavel to Dewar, 30/10/1900 and 21/11/1900 in *RIA* MM 1/1/1901; Scott to Dewar, 29/1/1901 in MM 25/2/1901; Petavel to Dewar, 1/12/1903 in MM 1/12/1903.
15 *RIA* MC DFRL, 7/12/1910.

16 Scott to the Managers, 28/11/1910. *RIA* MM 5/12/1910. The whole of the ensuing legal correspondence is attached to these minutes. Ludwig Mond had died in December 1909.
17 Professors at the RI were appointed for three years at a time.
18 Scott to the Managers, 1/2/1911.
19 *RIA* MM 5/12/1910, and further corespondence in *RIA* DC 1/2.
20 Strutt, see Chapter 3, note 14, p. 262.
21 J.J. Thomson to Scott, 20/3/1912, *RIA* DC 1/73.
22 *RIA* MM 3/7/1911, 29/1/1912, 5/12/1912, 3/2/1913, 5/5/1913, 3/11/1913, 6/4/1914, 8/6/1914, 2/11/1914, and MC DFRL 27/2/1911, 29/1/1912, 26/2/1912, 7/1/1913, 27/10/1913 report some of the complicated legal arguments and attempts at mediation that marked the long drawn-out course of the case.
23 Tilden to Scott, 7/5/1912, Ramsay to Scott, 19/5/1912, *RIA* DC 2/25 and 24.
24 *RIA* MM 2/2/1914, 2/11/1914. The phrase 'improvements in chemical science' is in the deed of endowment of the Fullerian Professorship.

Chapter 11 Decline

1 J. Dewar, President's Address, *Rep. BAAS* **72** (1902) 3–50; *Nature* **66** (1902) 462–77; *CP* 207.
2 Planck's paper 'On the theory of the energy distribution law of the normal spectrum' has been translated into English by D. ter Haar, *The Old Quantum Theory* (Pergamon Press, Oxford, 1964) pp. 79–81. Much has been written on the early days of quantum theory, see, for example, Kuhn, Chapter 8, note 138.
3 J. Larmor, 'On the application of the method of entropy to radiant energy', *Rep. BAAS* **72** (1902) 546 (abstract only); 'Radiation, theory of', *Emcyclopaedia Britannica*, 10th edn (1902).
4 Lord Kelvin, 'Nineteenth century clouds over the dynamical theory of heat', *PRI* **16** (1899–1901) 363–97.
5 Kelvin to Dewar, 6/6/1903 and 23/7/1903, *RIA* D II a(ii)/5 and 6. See also Chapter 8, notes 139–41.
6 The law of corresponding states implies a similar 'scaling' relation between the intermolecular potentials of the 'corresponding' substances. Kamerlingh Onnes's appreciation of this point became valuable later in the twentieth century for understanding the theory of liquids and liquid mixtures; see J.S. Rowlinson, *Cohesion: A Scientific History of Intermolecular Forces* (CUP, 2002) pp. 186–9, 250–1.
7 Kelvin to Dewar, undated, *RIA* D II a(ii)/10.
8 *RIA* MM 7/5/1902, 8/5/1911.
9 E. Crawford, J.L. Heilbron and R. Ullrich, *The Nobel Population, 1901–1937* (Office for History of Science and Technology, University of California, Berkeley, CA, 1987).
10 G.M. Prescott, 'Forging identity: the Royal Institution's visual collections', chap. 4 of *James*; *RIA* MM 7/5/1906. An analysis of the alleged audience in this picture has been made by Caroe, see Chapter 6, note 1, pp. 98–100.
11 H. Irving, 'Acting: an art', *PRI* **14** (1893–5) 419–28; *The Times*, 2/2/1895.

12 Ellen Terry to Dewar, undated, *RIA* D II c/12.
13 Dickson to Dewar, 8/3/1907, *RIA* D I e/44. I thank Jonathan Orr-Ewing for information on the membership of the Carlton Clubs in London.
14 *The Times*, 5/5/1880.
15 Armstrong, see Chapter 2, note 1.
16 J.G. McKendrick to H.E. Armstrong, 21/7/1923, *RIA* DB 11/1/3.
17 A.H. Savage Landor, undated manuscript, *RIA* DB 11/1/31.
18 Dickson, see Chapter 3, note 18. *RIA* DB 11/1/12a.
19 Green, see Chapter 9, note 40.
20 Cory, see Chapter 6, note 51.
21 Knott, see Chapter 2, note 3, p. 51.
22 *The Times*, 19/1/1885.
23 A.J. Berry and E.A. Moelwyn-Hughes, 'Chemistry at Cambridge from 1901 to 1910', *PCS* (1963) 357–63.
24 *The Times*, 1/5/1894, 18/12/1894.
25 T.A. Walker, *Peterhouse* (Heffer, Cambridge, 1935) pp. 125, 143. I am indebted to Roger Lovatt, the Archivist of Peterhouse, for the information that there is nothing in the minutes of the Governing Body that indicates any serious disagreement between Dewar and the College. When Thomas Walker had compiled his biographical register of the admissions to Peterhouse from 1615 to 1911 it was Dewar who undertook to pay 'the entire costs' of its publication in 1912, see *Peterhouse*, p. ix.
26 Pippard, see Chapter 4, note 6.
27 Tanner, see Chapter 3, note 1.
28 *RIA* MM 6/11/1916.
29 *The Times*, 27/3/1916.
30 K.S. Murray to RI Assistant Secretary, 1/5/1923, *RIA* DE 12/8/1; MM 4/6/1923.
31 Brock, see Chapter 2, note 40, pp. 451–6.
32 J. Dewar, 'Light reactions at low temperatures', *PRI* **19** (1910) 921–8; *CP* 235.
33 Young to Crichton Browne, 7/12/1912, *RIA* DC 1/1/1
34 Dewar's writing is not clear but Brock dates this letter to the 7th which is the Saturday.
35 Crookes to Siemens, 5/12/1913, 20/12/1913; Crookes to Young, 14/3/1914, *RIA* General Correspondence, CG 1r/3, 1r/5, 1r/7; MM 14/4/1913, 5/5/1913, 4/5/1914, 2/11/1914, 7/12/1014, the last of which contains the police report. S. Forgan, '… Some reflections on two hundred years of institutional history', chap. 2 of *James*.
36 Crookes to Meldola, 3/12/1912, from Meldola's papers at Imperial College, as cited by Brock, see Chapter 2, note 40.
37 Gordon to Crichton Browne, 9/4/1913, *RIA* DC 1/8.
38 Dewar to Armstrong, 20/2/1913, 23/5/1913, Imperial College Archives, C 288 and 292.
39 *RIA* MM 1/3/1915. After the death of his wealthy uncle he took the name Grove-Hills. His mother was the daughter of William Grove, the judge who is remembered for the Grove electric cell, which became an e.m.f. standard.
40 *RSA*, Council Minutes, 20/5/1915, 1/7/1915, 29/6/16.
41 *RSA*, Council Minutes, 5/7/1917; Council Documents 32.
42 J.J. Thomson to Arthur Schuster, 25/11/1917, cited by Rayleigh, see Chapter 3, note 35, pp. 194–5.
43 *RIA* MM 3/5/1915.

44 *RIA* D VII a/1, 7/6/1915.
45 *RIA* D VII a/2 to a/12.
46 *RIA* MM 16/6/1915.
47 *RIA* MM 16/5/1916, 6/6/1916; D VII a/13-64. The last of these files contains a printed copy of 'Notes for Managers by Professor Dewar' on this issue.
48 *RIA* MM 16/11/1916. The three men were Ansdell, Lennox and Heath.
49 *RIA* MM 6/11/1916.
50 *RIA* MM 4/6/1917, 2/7/1917.
51 *RIA* MM 5/2/1917.
52 Young to Gluud, 28/8/1919, *RIA* DE 12/9/3.
53 Lord Rayleigh, 'Foam' and 'Some aspects of photography' *PRI* **13** (1890) 85–97 and 249–54. A.S.C. Lawrence (1902–71), *Symp. Far. Soc.* **5** (1971) 8. He was known to my generation of research students as 'Soapy Lawrence'.
54 *RIA* D II d/3. Dewar's most substantial contribution to the field was described at length in his last Friday Evening Discourse on 19 January 1923, 'Soap films as detectors: stream lines and sound', *PRI* **24** (1923) 197–259.
55 McLennan to Dewar, 23/7/1921, Dewar to McLennan, undated, *RIA* D II c/5 and 6, with further letters, c/7 to c/12.
56 *Who was Who: 1916–1928* (Black, London, 1929).
57 *RIA* MM 3/12/1888: J. Dewar, 'Problems of the atmosphere', *PRI* **17** (1902) 223–8; 'Cloudland studies', *PRI* **23** (1921) 412–40; *CP* 205, 252. The 'charcoal thermoscope' measured heat by the volume of gas released by charcoal at low temperature when it absorbed radiant energy. Dewar described this device in his Friday Evening Discourse of 16 January 1920, 'Low temperature studies', *PRI* **23** (1920) 245–82, *CP* 251.
58 Will of James Dewar, dated 8 July 1922, and Codicil, dated 18 December 1922; proved, 16 May 1923; CGPLA, England and Wales. The accounts of the proceedings during his last days are not mutually consistent. The week around Dewar's death is described by Ralph Cory in his Friday Evening Discourse, note 20, above. He describes a short funeral service for the staff at the RI conducted by [William Hartley] Carnegie, a Canon at Westminster Abbey, and a former Principal of King's College, London, whom he describes as an 'old friend' of Dewar's, but he gives no date for this service. Arthur Keith says that a service was conducted on the Saturday morning by the Bishop of Gloucester; *An Autobiography* (Watts, London, 1950) pp. 460–1. *The Times* (2/4/1923) names this man as the Bishop of Worcester, Ernest Harold Pearce, *ODNB*, who is a more plausible candidate, sharing many points of possible contact with Dewar. He had recently been elected an Honorary Fellow of Peterhouse, he was a keen musician, being Chaplain of the Musicians' Company, he was the ecclesiastical correspondent of *The Times* (and so a colleague of Dewar's friend and apologist Munro Ross) and he was an advocate of cremation. Landor was at the Crematorium but describes the group there ambiguously: 'the only people present included the Duke of Northumberland, Sir Arthur Keith, Sir Alexander Mackenzie, Sir James Crichton Browne and myself'; *RIA* DB 11/1/31. This document is undated but must be almost contemporary with the event since Landor died in Florence the next year.
59 *The Times*, 9/1/1935, 12/1/1935, 21/2/1935, 5/3/1935, 29/3/1935 and 30/3/1935.
60 A.R.J.P. Ubbelohde (1907–88) F.J. Weinberg, *BMFRS* **35** (1990) 383–402.
61 *RIA* MM 6/2/1893 and 8/5/1899.

62 *RIA* DB 11/4, 24/1/1921.
63 G.M. Prescott, 'Forging Identity', chap. 4, pp. 81–4 of *James*; *PRI* **23** (1921) 441–7.
64 Addresses on 9/4/1923 in the RI; *PRI* **14** (1924) 81–6.
65 *RIA* MM DFRL 25/4/1923; Keith, note 58; Caroe, see Chapter 6, note 1, 'The change in outlook', chap. 6, and J. Hughes, 'Craftsmanship and Social Service: W.H. Bragg and the modern Royal Institution', chap. 10 of *James*.

Name Index

An entry of the form **7**, *11* denotes that biographical or similar information is to be found in note 11 of Chapter 7.

Subject Index

For Product Safety Concerns and Information please contact our EU representative GPSR@taylorandfrancis.com Taylor & Francis Verlag GmbH, Kaufingerstraße 24, 80331 München, Germany